Bicycling to Utc

Bicycling to Utopia

Essays on Science and Technology

Edited by
P. DAY
and
C. R. A. CATLOW

Oxford New York Tokyo
OXFORD UNIVERSITY PRESS
THE ROYAL INSTITUTION
1995

Oxford University Press, Walton Street, Oxford OX2 6DP

Oxford New York
Athens Auckland Bangkok Bombay
Calcutta Cape Town Dar es Salaam Delhi
Florence Hong Kong Istanbul Karachi
Kuala Lumpur Madras Madrid Melbourne
Mexico City Nairobi Paris Singapore
Taipei Tokyo Toronto

and associated companies in
Berlin Ibadan

Oxford is a trade mark of Oxford University Press

Published in the United States
by Oxford University Press Inc., New York

© The Royal Institution of Great Britain, 1995

All rights reserved. No part of this publication may be reproduced, stored in a retrieval system, or transmitted, in any form or by any means, without the prior permission in writing of Oxford University Press. Within the UK, exceptions are allowed in respect of any fair dealing for the purpose of research or private study, or criticism or review, as permitted under the Copyright, Designs and Patents Act, 1988, or in the case of reprographic reproduction in accordance with the terms of licences issued by the Copyright Licensing Agency. Enquiries concerning reproduction outside those terms and in other countries should be sent to the Rights Department, Oxford University Press, at the address above.

This book is sold subject to the condition that it shall not, by way of trade or otherwise, be lent, re-sold, hired out, or otherwise circulated without the publisher's prior consent in any form of binding or cover other than that in which it is published and without a similar condition including this condition being imposed on the subsequent purchaser.

A catalogue record for this book is available from the British Library

Library of Congress Cataloging in Publication Data available
ISBN 0 19 855895 3

Typeset by Footnote Graphics, Warminster, Wilts
Printed in Great Britain by
Biddles Ltd, Guildford and King's Lynn

PREFACE

When I was growing up as a boy in a small village in Kent in the 1940s, Friday night was bath night. At the Royal Institution, since 1826, we have known differently; Friday night is Discourse night. In fact, the database of Discourse speakers and titles, maintained now on computer, informs me that since the first Friday Evening Discourse given by Michael Faraday on 'Caoutchouc' on 3 February 1826, there have been no fewer than 3349. Even among such a large number, the Discourse given by the Right Honourable William Waldegrave is an especially notable one, for two reasons. First, it marks the beginning of the National Week of Science, Engineering and Technology, an event instigated by the Office of Science and Technology as part of the campaign to enhance public appreciation of these matters, foreshadowed in last year's Government White Paper *Realising Our Potential*. Second, to the best of my belief, it is the first time in the 168-year history of the Friday Evening Discourses that a serving member of the Cabinet, and Minister responsible for Science, has given one.

The Friday Evening Discourses were arguably the first sustained endeavour anywhere in the world to help the public towards a better understanding of the scientific approach to our world, although the Royal Institution's efforts in this direction go back even further, to Humphry Davy's popular lectures on chemistry. One of the lesser known facts of London life is that the first ever street to be officially designated 'one way' is the one in which the Royal Institution is situated. The reason why it was so designated is perhaps even less well known: it is a tribute to the huge popular success enjoyed by Davy's lectures. Mayfair high society came in such numbers in their carriages that the street was blocked, and a traffic control system had to be installed.

At the beginning of the nineteenth century, science was considered an integral part of the nation's cultural life and, for example, Humphry Davy was a close friend of both Wordsworth and Coleridge, both of whom stood at the rostrum in our Lecture Theatre. Indeed, in a flight of poetic fancy uniting art and science, Coleridge wrote:

> Water and flame, the diamond, the charcoal. .. are convoked and fraternised by the theory of the chemistry. ... It is the sense of a principle of connection given by the mind, and sanctioned by the correspondency of nature. ... If

> in a Shakespeare we find nature idealised into poetry, through the creative power of a profound yet observant meditation, so through the meditative observation of a Davy. ... we find poetry, as it were, substantiated and realized in nature: yea, nature itself disclosed to us .... as at once the poet and the poem!

Would it were so today. Indeed, during the consultation that led up to last year's White Paper, many of us urged that improved public appreciation of the role of science, engineering and technology in our society was an essential underpinning of any national strategy in these matters.

And so, here we are, at the beginning of what has to be a sustained and continuous programme of endeavour, to bring science back to the centre of the stage: why, I always think, to quote the founder of the Salvation Army, should the Devil have all the best tunes?

London P.D.
March 1995

CONTENTS

PLATES

Plates fall between pages 52 and 53, 84 and 85, and 126 and 127.

CONTRIBUTORS

M. D. Archer
The Old Vicarage,
Grantchester,
Cambridge CB3 9ND

D. T. Clark
Scientific Director,
Research Unit for Surfaces,
Transforms, and Interfaces,
Daresbury, Warrington,
Cheshire WA4 4AD

Iwao Fujimasa
Professor, Research Center for
Advanced Science and
Technology,
University of Tokyo,
4-6-1 Komaba,
Meguro-ku, Tokyo 153,
Japan

Tim Harper
Statoil-Forskningssenteret,
(Rotvou), Postuttak,
N-70004 Trondheim,
Norway

J. S. Jones
Professor of Genetics,
Galton Laboratory,
University College London,
Gower Street,
London WC1E 6BT

Alan Maynard
Professor, Centre for Health
Economics,
Heslington,
York YO1 5DD

Michael O'Shea
Professor, Sussex Centre for
Neuroscience,
School of Biological Sciences,
University of Sussex,
Brighton BN1 9QG

William Waldegrave, MP
Ministry of Agriculture, Fisheries
and Food,
Whitehall Place,
London SW1A 2HH

Malcolm B. Wilkins
Regius Professor of Botany,
Institute of Biomedical and Life
Sciences,
Bower Building,
University of Glasgow
Glasgow G12 8QQ

National Week of Science, Engineering, and Technology (SET⁷)

The Rt Hon WILLIAM WALDEGRAVE MP

Chancellor of the Duchy of Lancaster
Minister of Public Service and Science

It is a humbling thing for an ex-classicist to stand at the podium theatre of the Royal Institution, conscious of the formidable ghosts of Faraday and Davy looking sceptically down on him, feeling around him the presence of the Braggs, knowing that the lively eye of George Porter may well be upon what he says. Nonetheless, it was the greatest of the Roman poets, writing of the only man ever to compose a wholly successful epic poem about science, Lucretius, who gave us the words which are the first celebration of the scientist's destiny:

> Felix, qui potuit rerum cognoscere causas

Latin is a better language than English for the lapidary comment: *felix* means happy, and lucky, and blessed—all attached to the man (in this case) who can understand the causes of 'things', for that is how we translate the word *res*, but not with the trivial resonance of English. But at least Virgil gave me the tribute to lay at the feet of this great Institution.

On the day on which the Discourse was given, 18 March 1994, we launched, with the splendid help of the British Association, SET⁷, Britain's first Science, Engineering, and Technology Week. It was thoroughly fitting that the launch should be formalized at the Royal Institution, where for nearly 200 years, the tasks of scientific discovery and of the explanation of science have been pursued jointly. To this day, first rate science is done there; to this day, the Royal Institution carries the banner for the public understanding of science which is the central objective of SET⁷.

In this article, I want to spread the net a little wider, and discuss the place of science in our society at large.

The Government will shortly be publishing the first annual Forward Look document, promised in my White Paper *Realising Our Potential*, which will, amongst other things, give a summary overview of where we think British science and technology stands in 1994. It will inevitably be a somewhat rough and ready picture; we hope to refine it steadily each year as the networks we are establishing with universities, industry, government departments, and individual scientists and engineers become stronger. But I am giving away no secrets when I tell you that, notwithstanding all the problems (and we will not ignore them in our account), British science remains extremely strong. It is first or second in Europe in virtually every important area, if you are to believe the *Science Citation Index*, and overall (according to that measure), second only to the United States; high in the Nobel Laureates scoreboard; and a centre for a disproportionate number of famous research laboratories. In short, our science, unlike a good many things in this country, is not part of a glorious past, it is part of a present of very great vitality.

I want to address two questions here. First, is science good for a society? And second, since I shall give an affirmative answer to the first question, what do we need to do to maintain our scientific strength?

Within the walls of the Royal Institution, the answer to the question 'Is science good for society?' may seem self-evident. But to many thoughtful people outside these walls, the answer is not so obvious. Some see science, and the methods of science, as systematically destructive of everything which makes life worth living. 'Thanks to Newton,' says Brian Appleyard, 'we cannot discover goodness in the mechanics of the heavens, thanks to Darwin we cannot find it in the phenomenon of life, and thanks to Freud we cannot find it in ourselves'.(1) This bleak view has often found powerful expression in art and literature, from Richard Jefferies' grim vision in the last century of a future London sinking into a mire of pollution and self-destruction, via the nightmares of *Brave New World*, to the powerful genre of apocalyptic literature of the post-Hiroshima world. Skilful politicians have usually judged that the populist vote lies with those who are anxious about where science is taking us. 'Is man an ape or an angel?' asked Disraeli, apropos of the battles which surrounded the publication of *Origin of Species*, 'Now, I am on the side of the angels'. Winston Churchill advised that the scientist should be 'on tap, not on top'; while by far the easiest way for someone with a smattering of scientific education to become a media guru is to set himself up as a purveyor (for a fee) of scientific scare stories to the popular media, describing himself as Dr X, well-known expert in food/nuclear/ecological safety. The doctorate, incidentally, can be in anything from librarianship to physical education.

But it is no good just complaining. The arguments have to be confronted and answered. They fall into two kinds. First, let us consider the argument summarized by Appleyard, that since scientific results are always provisional, subject to revision, and inherently uncertain, the scientific method inevitably undermines the necessary moral and spiritual certainties lacking which life descends into a meaningless morass without truth, faith or belief.

This is the least serious objection to science. I believe it is simply wrong, or rather it is a worry that can easily be stilled. Since the beginning of articulated thought on these matters, roughly in the sixth century BC, so far as we know, in Asia Minor and Greece, people have played with the paradoxes which derive from the fact that there are different ways of describing things for different purposes. A table may be, a mixture of four elements, or a flux of atoms, or an indescribable mess of waves, particles, spins, charms, Higgs' bosons, and who knows what, but for the purposes of everyday life it remains a table. Right and wrong may be illuminated by a hundred great philosophers (though, in the end, they mostly add rather little to Aristotle) but the interesting thing is that, whatever conflicting language we use to describe the source of the authority of those concepts, we end up using the words without much difficulty, and translating them from English to Japanese via all stations in between. For those who do not share the atheism of some great scientists (and non-scientists too), it is really not very difficult to describe the different sense in which I may say I believe in God from that in which I say I believe in the existence of pulsars.

Indeed, some of those who seem in my experience to have the clearest perception of moral values and the most lucid spiritual beliefs have been amongst the great scientists I have had the luck to know. Indeed, they very often seem to me more effective upholders of social and moral order, and more receptive of beauty in art and nature, than many who spend their lives explicitly defending these allegedly non-scientific values. Nor do I believe that they are being duped into destroying, by their science, the moral foundations of their non-scientific beliefs. To believe that they are is to commit what, in my days, was described by Gilbert Ryle (though it is really in Aristotle) as a category mistake. Indeed it is possible to argue, as Plato effectively does, that the search for knowledge itself, whatever the prizes available, always has a disinterested element to it, and therefore a moral dimension. Do not misunderstand me: I know there are plenty of rascals amongst scientists, too.

The less grand argument, the populist argument, against science, is more difficult to deal with. This runs as follows. The argument is that mankind's inherent weaknesses—greed, aggression, short-sightedness, and so on—mean that we are bound to misuse any powers we have, and that the more powers we have, the more dangerous we are to ourselves and to other species on the planet.

There is of course a truth in this. A terrorist of Conrad's day with a gunpowder bomb could kill a few people and often himself. A terrorist of today with Semtex can kill a lot of people from a safe distance. What could a terrorist of tomorrow with a nuclear device or a biological weapon do? We can now do things which could affect the future of the species, if not the ecosphere. We can, at lower levels, make designer drugs and send vivid pornography into every home. All these greater dangers and temptations derive from the greater powers which our own intelligence, deployed in science and technology, make available to us.

You will find this story in the Book of Genesis: it is the story of man's Fall. The rest of that volume is one part of the answer. It is true that the first of our ancestors who used a tool also invented the first potential weapon, because a weapon is simply a tool applied with a different intention. But you cannot separate the danger of the misuse of the tool, or of science, from the nature of our species. You cannot say that we will abandon our curiosity and our inventiveness because of the dangers of the misuse of what we discover, because curiosity and inventiveness are two of our fundamental defining characteristics as a species—perhaps *the* defining characteristics. Are we not called '*Homo sapiens*'? It is possible to imagine a society where the organized fear of new knowledge is allowed to suppress enquiry—there have been some, here and there, and their remains are of interest to archaeologists. Imagine trying to go down that route now. No natural ancient equilibrium would reassert itself; no rural idyll grows amongst the deserted laboratories and burnt out libraries. Lovelock has said all this well; the equilibrium we need now has to be the product of our own conscious, intelligent purpose. Going back is not an option.

But the argument is also a positive one. To continue for a moment with the language of Christianity, man may have fallen, but he has the hope of salvation. That is, we are dangerous, but we are not *only* dangerous. We are also the most successful species known to the planet, not yet in longevity but in our capacity to repair damage whether self-inflicted or externally derived; we are the only species to have left the planet, albeit as yet only on a very short journey; we are the only species which can analyse our predicament and act on it. Less grandly, we are the only species which can make our circumstances of living more pleasant by conscious thought: we can cure some of our pain, communicate over great distances, keep ourselves warm or cold, and all the rest. We are also the only species with a sense of the sublime, of things above ourselves, of beauty, and of art. The first tool was potentially the first weapon, but it was also potentially Rembrandt's brush or Einstein's pencil. So the answer to my first big question, 'is science good for a society?, is a positive one; we can both dismiss Appleyard's dehumanizing claim and, while recognizing that the populist's warning that science can be the servant of the dark side of our humanity has some force, conclude that it is also part of the highest expression of the glorious aspect of being human. For every misuse of our intelligence we can imagine, or demonstrate, there are a hundred positive uses—including increasing our skill in learning how to moderate our dangerous side.

Any society which is truly vigorous is so by integrating and ordering not the lowest common denominator of our nature but the highest factor: a society organized to allow and celebrate the creative spirit in science will find itself also productive of the other forms of creativity which make like worth living. They are all bound in together, and always have been. The societies—national or more local—where the bursts of scientific energy occur which astound the historians also span the other arts too. The rule runs from fifth-century Athens to Renaissance Italy to Queen Anne's England, and on down to the modern world. A society which is alive looks forward, has energy, will not take no for an answer: all those characteristics, given a little luck, throw up the scientists, philosophers, painters, musicians, and the rest. Often if you look at those societies it seems somewhat of an accident who turns into what: would Bach in a different family write Leibniz and vice versa? The Jewish violinists and physicists get inextricably muddled up; and it is not difficult to imagine Newton designing great classical buildings. And then there is Leonardo!

At one level, therefore, I am profoundly a Leavisite. There are not two cultures, but one. What there is, however, is a great danger of over-specialization, which can be a cause of sterility and decay in a vigorous society. What Snow said of the ignorance by his literary friends of the Second Law of Thermodynamics can in fact be duplicated a thousand-fold within each of his divisions.

We do not have a simple 'Cold War' confrontation between science and arts; we have a Balkanization of all culture into thousands of fragments. The expert on the Athenian Tribute lists may be as ignorant of romantic German painting as he is of the human genome project; the numbers theory man may make a less good attempt at describing how an internal combustion engine works than does the econometrician, whose graphs, in turn, may mean less than nothing to the politician in charge of the Finance Ministry. This Balkanization is the real danger and it is a real source of weakness. In the list of super-creative societies I mentioned, the unity of intellectual life is rather obvious: in fifth-century Athens the scientist *knew* the historian who knew the philosopher and the play-wright. So too in Renaissance Florence and Milan; so too in London in Wren's and Newton's day, and so on.

Our danger now is that, although communication technologies give us the methods of doing it, we have not learnt to use them to recreate in modern form the vital interactions and cross-fertilizations which through history seem to spark the engine of those societies at which we look back with most envy.

You will see that I have edged my way towards some suggestions about the answer to my second big question. Given that vigorous science is one of the characteristics of a healthy society, how do we ensure our scientific strength? You will see that I think that one essential is cross-disciplinary fertilization. We should be careful of the health of the classical institution invented for this purpose, the university. But buildings and physical propinquity are not enough if esteem is given only to greater esoteric specialization. A college full of non-communicating experts is not what I mean by a university. A good decanter of port after dinner will help, but even that is not enough. We have consciously to address the question, of equipping ourselves with enough language in common to be able to communicate across disciplines, whether scientific or not.

My conclusion is not just for scientists—in fact it is by definition relevant, if I am right, to all those who are engaged in the essential human activity of creation and innovation, whether in the arts or technology or politics or anywhere else. How do we do it? How do we learn to talk to each other again not just at the commonplace level, but in order to explain to one another the difficult parts of our subjects? If we do not manage it, we will I fear so divide the stream that the rivulets will dry up in the desert. Well, there are some practical, do-able things we could set our hand to. We could seek to give all scientists and engineers a wider grounding in history, philosophy, or aesthetics. We could ensure that our arts graduates are not innumerate and know a little at least of the history of science. This can be done, because it is done in some places already.

Then, we could take care to ensure, in order to balance the rival claims of institutional specialization, that there are a decent number of very strong, multi-faculty universities, and should beware of the temptation, in a geographically small country, to allow the mobility available to our students to take us too near to specialist universities. There will be a need for some, but not all must go that way. Here are things that governments and educationalists, over time, carefully, can do. But what of the wider culture? Is there anything practical that can be done?

I believe there is, and in its way the SET[7] week embodies my belief that it is worth trying. We need to reward, with esteem but not only with esteem, those who are willing to cross frontiers. This Institution has been at this sort of work for nearly two hundred years. We can deploy money to support, in the modern idiom, similar work through the media. We can, from the science side, make a little evangelism a necessary part of the job for those we back to do research. We can work with those in industry who live by selling the products of science—and that is, to all intents and purposes, everybody in industry—to emphasize the utilitarian arguments, the arguments based on the pressing need for new products to sell and new processes to enhance the quality of life. We can, whatever our discipline, seek to reunify a culture in danger of fragmentation.

If we can do that, we will have created conditions for the preservation of the strength of our science, but not just of our science. The money? Well indeed, proper government funding of science—particularly but not only of basic science—is vital. But ask yourself whether the funding will be safer in a democracy where the university physicist feels him or herself part of the same culture as the economics graduate reading the news and both of them understand somewhat more the engineer running the railway—that is, where more people understand the importance of science; or in a world where science holds the place of Eleusinian Mysteries? I think the answer is obvious.

WILLIAM WALDEGRAVE, MP

Born 1946, educated at Eton and Corpus Christi College, Oxford, where he was President of both the Oxford Union and The Oxford University Conservative Association. He was a Kennedy Fellow at Harvard University. He was Fellow of All Souls College, Oxford 1971–86, member of the Central Policy Review staff, Cabinet Office 1971–73, and Political staff of 10 Downing Street 1973–4. He has held numerous other positions in both Opposition and Government after becoming Conservative MP for Bristol West in 1979, including four Ministerial posts. Secretary of State for Health 1990–2. He became an Honorary Fellow of Corpus Christi College in 1991, and Chancellor of the Duchy of Lancaster in 1992.

Bicycling to Utopia

J. S. JONES

Genetics and fear seem to go together. Freudians might argue that this is because both have their roots in sex. True or not, there is something about the study of inheritance that touches on our deepest anxieties.

The worries today are about our individual fate—coded, for many, into the genes, with two people of every three dying for reasons connected with the genes they carry—or that of our children. These fears are rational even if we can do nothing about them. In the earliest days of human genetics, though, the concern was a deeper one.

In the twentieth century physics and chemistry have produced so many different ways of destroying the human race that there seems almost no point in worrying about the biological future. Such nonchalance is new. Human genetics began with Francis Galton, Charles Darwin's cousin, founder of the Galton Laboratory at University College London (UCL), where I now work. He argued in his 1869 book, *Hereditary Genius*, that those of innate merit—the geniuses of his title—were having too few children, and that as a result the human race was on the edge of decline. Something, he felt, must be done. His views led indirectly to some of the greatest disasters of the next hundred years.

Galton was a strange and in many ways unlikeable man. Every day at UCL I walk past a collection of relics of his life. They include a battered book of fingerprints (including those of the Prime Minister William Ewart Gladstone), an old copy of *The Times*, and a brass counting gadget. Each is a reminder of his extraordinary range of interests. He revolutionized detective work by introducing the idea of using fingerprints as a clue of individual identity. What is more, Galton was the first person to publish a weather map and the only one to have made a beauty map of Britain, based on secretly grading the local women on a scale of one to five (the low point, incidentally, was in Aberdeen).

Much of his work had to do with measuring human quality. Insofar as genetics does the same thing, he was the founder of the scientific study of human inheritance (although, of course, he knew nothing of Mendel). His concerns about the future of human evolution imprinted themselves on many of his descendents. A surprising number of Utopian novels trace their visions of the future to Galton and Darwin. Aldous Huxley's *Brave New World* owes much to his grandfather Thomas Henry Huxley, 'Darwin's bulldog'. H. G. Wells—whose Utopia appeared in *The Shape of Things to Come*—himself wrote a textbook on evolution with Aldous Huxley's brother Julian, and George Bernard Shaw, author of *Back to Methuselah*, appeared in public with Galton.

Since Galton we have learned an enormous amount about human evolution. We can now reconstruct much of our past, not from fossils (although they are the best of all evidence that evolution has in fact taken place) but from the records of ancestry preserved in our genes. The process of evolution, too, is now largely understood. Perhaps, then, we are for the first time in a position to replace the unfocused fears of earlier generations with some real predictions about the future of human evolution.

The theory of evolution in its modern guise is remarkably close to that proposed by Darwin, with the addition of a mechanism of inheritance of which he was ignorant. The Darwinian machine has three parts. The first is a means of generating inherited diversity—mutation, as it is known. The second is a filter, which sorts out mutations on the basis of quality. Quality, in this context, is easily measured: it depends only on the ability to reproduce. Those variants which are more successful in copying themselves survive and in time become more common. Those which are less effective die out. This is called **natural selection**.

Finally, there is an important (but often forgotten) aspect of evolution: very often, it takes place at random. The accidents of history mean that one gene prevails at the expense of another, even though they do not differ in quality. This process, **genetic drift**, is particularly important in small populations. As, until a mere ten thousand years ago with the origin of agriculture, there were no more people in the world than there are in London today, a great deal of human evolution must have happened by chance.

There is no reason to suppose that the nature of the Darwinian machine will change: it has, after all, been at work since the origin of life three thousand million years ago. If we can predict what will happen to mutation, selection, and genetic drift, we will be in a position to make informed guesses about the future of human evolution.

Many people are convinced that some of the 'benefits' of modern civilization—radiation, chemicals, and the like—will increase the damage to our DNA. It is certainly true that industrial chemicals and nuclear radiation increase the rate of mutation. However, before worrying about their effects on future generations it is helpful to look at them in the context of natural agents of mutation which have always existed.

The most cynical experiment in the history of genetics took place on 6 August 1945. On that date an atomic bomb was dropped on Hiroshima. Two days later, another fell on Nagasaki. They ended the war with Japan—as the Emperor pointed out, in his first and somewhat understated broadcast to his people, 'The situation is not necessarily developing in Japan's favour.' Soon afterwards, the Americans sent a team of scientists into the devastated cities. It set out to test whether the children of irradiated survivors of the bombs carried any new genetic damage. The search centred on a massive survey of the structure of proteins using a technique known as electrophoresis.

The Final Report appeared in 1991. Children were divided into two groups: those whose parents were less that 2.5 km from the burst and those further away when the bomb fell. Every survivor was interviewed to try to find out where they had been at the time of the explosion, whether they had been sheltered by buildings, and how they had been standing in relation to the source of the radiation. From this it was possible to estimate the dose which each had received.

More than a million genes coding for proteins were looked at. They were tested to see if there were any changes in structure in children as compared with their parents. Three mutations were found in children whose parents were in the cities when the bombs fell and three in the children of those outside the fallout zone. These people have probably had more radiation than any others in history, and yet the rate of DNA damage does not seem to have increased much.

Some populations have been exposed to radiation for many years. The biggest source in Britain is radon gas, which leaks from granite. People living in granite houses in Cornwall may be exposed to more excess radiation than those working in nuclear power stations. Those at risk are advised to install fans to stop the gas accumulating. There are other sources of radiation (such as medical X-rays and even luminous watches) but for most people these involve very small doses. Regular air travellers also suffer a noticeable increase in exposure to radiation.

Some chemicals are potent mutagens. They are a much more universal threat than is radiation. However, once again, it is important to put their threat into the context of what is there already. A recent survey of the mutagenic effects of pesticides shows this clearly.

There are two types of pesticides. By far the most abundant are natural chemicals produced by plants as a defence against their insect enemies.

In addition there are the various products of the chemical industry which have added a few simple weapons to this natural armoury.

Testing the ability of both to cause mutation (in this case in body cells rather than in the sperm and egg) shows that some are indeed potent agents of biological damage. However, the list makes chastening reading. Taking into account the average daily exposure of a typical westerner, nearly all the most damaging agents come from perfectly natural sources.

Top of the list is caffeine. Perhaps, given that it is taken for pleasure, that is not unreasonable. Next, though, is lettuce—which has, in its bitter heart, a powerful insecticide. As if that was not enough, natural orange juice (flakes of peel included) comes next. Artificial pesticides, such as DDT, are far less damaging.

One of the most powerful of all chemical mutagens is aflatoxin, found in the moulds growing on badly stored foods. Even in the past fifty years the exposure of westerners to this substance has dropped dramatically and, with this, so has the number of new mutations. There is no need to fear a nuclear or chemical nemesis awaiting our genes.

In fact, a more subtle transformation is having a dramatic effect on the mutation rate. In the western world at least, a change in the age at which people have children means that the number of new mutations will probably drop.

The rate of mutation goes up greatly with age. Most people now live for far longer than in earlier times. Mutation can hence take its toll on a higher proportion of the population. This is obvious when looking at such changes in the body cells, including those which give rise to cancer. The cancer epidemic in the modern world is largely confined to older people. A shift in the pattern of survival has had effects on genes as they reside in body cells.

Cells which give rise to sperm or egg are also exposed to the destructive effects of old age. Older parents are more likely to have genetically damaged children than are those who reproduce early. For some genes the rate is 30 times higher in parents over 40 than it is in those under 20. Any change in the age of reproduction will hence have an effect on the mutation rate. If the number of older parents goes up, there will be more mutations; if it decreases there will be fewer. Social progress has led to just such a shift. The general picture, worldwide, is simple and a little surprising. Until recently, mothers started having babies when they were themselves youthful and continued to have them until they were biologically unable to do so, perhaps 25 years later. As infant mortality drops there is less pressure to have children as an insurance against one's old age. The availability of contraception means that parents can choose to delay their first child—sometimes, as in middle class Britain, until their mid-twenties—but then complete their families quickly. This means that most people stop having children soon after they have started and so the number of older fathers and mothers goes down.

The effect is obvious in post-war Europe. In countries such as Switzerland, the proportion of mothers aged more than 35—the group most at risk of mutation—dropped from around 20 per cent in 1950 to less than five per cent in 1985 and is still falling. The effect is particularly striking in Ireland. The influence of the Church meant that until a few years ago the only effective means of birth control was self-denial. Most Irish people did not marry until their late twenties, or even later, and until recently there were more than twice as many mothers over 35 in Ireland than anywhere else in Europe. The number is now dropping rapidly (although it is still well above the European average). In Britain in the mid-1970s there was a slight reversal of the trend towards earlier reproduction, with the number of mothers over 35 increasing slightly from its low point of around one in 20.

All this means that there are fewer old parents than there have been for much of the evolutionary past. This is bound to have an effect on the mutation rate. Down's syndrome is 10 times more common in mothers over 45 than those under 21. It is three times more common in Pakistan (which has almost no family planning) than in Britain, largely because Pakistani mothers are older than their British equivalents. At the moment, at least, it looks as if the human mutation rate is on the way down. Whether this trend will continue is not known, but it does put fears about a new race of mutated monsters into context.

If there has been no real change in the supply of mutations, the fuel of evolution, what about the future of natural selection, its engine?

One concern which can immediately be dismissed is that somehow genetic technology will alter our genes for the worse. For the most frequent single-gene inherited diseases, such as sickle cell anaemia and cystic fibrosis (the commonest diseases among those of African and European ancestry respectively), two copies of the damaged gene are needed, one from each parent, before any harm is done. Modern medicine means that both illnesses can now be treated, and those affected may survive and have children. In contrast, new tests which show whether a fetus is at risk mean that the number born with such diseases is dropping as parents choose abortion.

It is hard to predict the balance between the two opposing forces. In some places, the availability of genetic tests has led to a dramatic drop in the number of children born with a damaged gene. In Ferrara, in Italy, for example, a test for a common local inherited blood disorder was introduced in the early 1970s. The number of children born with the condition has gone down by more than 90 per cent. However, a few of those born may now live for long enough to marry and pass on a copy of their own genes.

Because such inherited defects are individually so rare, any change in their ability to survive will have little effect on the immediate evolutionary future. There are, however, other ways of looking at the future of natural selection which suggest that, in the West at least, it is losing its force.

The history of one inherited character, the weight of babies, shows just how effective improved conditions can be in reducing the action of selection. Birth weight shows the advantage of being average. Not surprisingly, underweight babies survive less well than others. What is more remarkable is that babies heavier than average are also more likely to die in the first few weeks after birth. In the 1930s, about half of the babies who died in their first year were above or below the ideal weight. A difference of just one pound had a major effect on survival. Selection was at work against genes for high or low birth weight as it had been, no doubt, ever since our species began.

Now, it is disappearing. Only very underweight babies, or those much larger than average, are at risk. The intensity of natural selection went down by two-thirds between 1954 and 1985. What must once have been one of the strongest agents of selection (acting as it does before the age of reproduction) seems to be on the way out.

There are other, more subtle ways of looking at the future. Natural selection can only act on differences. If everyone survived to adulthood, found a partner, and had the same number of children, there would be almost no chance for selection to operate. Looking at changes in the patterns of birth and death reveals a lot about its actions.

In affluent countries, the differences between families in how many people survive have decreased. This means that there is less opportunity for natural selection. Ten thousand years ago, the struggle for existence really meant something. Skeletons from cave cemeteries show that few people lived to be more than twenty. If ancient fertility was anything like that found in modern tribal groups, each female had about eight children, most of whom died young. For nine-tenths of human evolution, society was like a village school, with lots of infants, plenty of teenagers, and a few—probably harassed—adult survivors. Almost every death was potential raw material for selection as it involved someone young enough to have a hope of passing on their genes.

Now, things have changed. Ninety-eight out of every hundred newborn British babies live to the age of 15 so that selection acting through childhood deaths (once its main mode of operation) has almost disappeared.

What about the last part of the Darwinian machine, evolution by accident? It has certainly had a dramatic effect on the genetic past. How it works is illustrated by the evolution of a simple and unimportant character, the surname. Just like a gene this is passed down the generations (albeit through only one sex) and, just like genes, it does odd things in small populations.

The Afrikaner population of South Africa has remarkably few surnames—Botha and De Clerk are just two of the twenty or so shared by a million Afrikaners. The reason for the small number of surnames is that only a few families (and their names) migrated from Europe, so that their descendants retain only a small sample of the names of The Netherlands. The founders brought their genes as well as their names, and this accident of history had a dramatic effect on the genes of their descendants.

One of the partners in the marriage of Gerrit Jansz and his wife Ariaantje Jacobs (who was one of a group of girls sent from a Rotterdam orphanage in the 1660s) must have carried a single copy of the gene for a form of porphyria. This disease is due to a failure in the synthesis of the red pigment of the blood. Some geneticists suggest that the madness of George III was due to a particularly virulent form of the disease. The symptoms vary from case to case. Sometimes light-sensitive chemicals are laid down in the skin. Here, they react with sunlight and can produce painful sores. In some forms of porphyria, hair grows on exposed areas. Sometimes the chemicals accumulate in the brain, leading to mental disorder. Part is excreted in the urine, giving a characteristic port-wine, almost blood-red, colour. Some claim that the origin of the werewolf legend—those creatures which come out only at night, howl, and drink blood—may lie in the porphyria gene.

The South African form is relatively mild but became important when barbiturates were used in the 1950s. Those with the gene suffered pain and delirium when they took these drugs. Porphyria is rare in Europe but thirty thousand Afrikaners carry it. In Johannesburg, there are more carriers than in the whole of The Netherlands. All descend from a single member of the small founding population which grew in numbers to produce today's Afrikaners. Because it is so common in one family line in South Africa, the disease is sometimes called van Roojen disease. A gene and a surname have thus become intimately associated and tell the same story about history. The evolution of this population has been has been driven by an accident of the past.

Just the same thing happens in any small and isolated group. In Switzerland, for example, everyone in a mountain village may have one surname, while the whole population in another a few miles away shares a different one. By chance and over the years within each hamlet there has been an accidental loss of names as some men have no sons. Different names take over in each place. The process is inevitable: go on for long enough in isolation and everyone will have the same surname, the same ancestor, and the same shared genes.

Finland, with its impenetrable forests, has many isolated and inbred populations, and many foci of inborn disease. The same thing happens in North America. The Jemez, in New Mexico, have an astonishing number of albinos because, in the distant past, one of their founders carried the albino gene; because of inbreeding the effects of this gene are now being manifested.

Social barriers may be just as effective as geographical ones. In Britain, many children of Pakistani immigrants marry among themselves. Nearly half are married to a cousin. Although only about one British birth in fifty is to such parents, they represent about five per cent of all inborn disease.

Just as is the case for mutation and natural selection, all this is changing. Without anyone noticing, we are in the middle of the most dramatic event in evolutionary history. It may mean that the human race is entering an era of unprecedented genetic good health—a biological Utopia reached by accident.

It all has to do with a change in mating patterns. During most of history, everyone more or less had to marry the girl (or boy) next door, because they had no choice. Society was based on small bands or isolated villages and marriages were within the group.

Now this pattern is shifting. An increase in mating outside the group is the most striking change in the recent biological past. The effect of this outbreeding on genetical health will outweigh anything that medicine is able to do.

Much of it arises, like so many of the biological events which shape the human condition, as a by-product of social change. Cities and transport both played a part. Both increase the pool of potential mates. The tens of thousands of surnames in New York show just how mixed up the world's population is becoming.

In Europe a century ago, marriages between relatives were still common. In the Aeolian Islands off the coast of Italy in the 1920s a quarter of marriages were between first or second cousins. This figure has dropped to about one in fifty. Another way of illustrating shifts in breeding pattern is to use a crude but effective measure of relatedness. All that one needs to know is how far apart husbands and wives were born. If they come from the same village they may well be relatives, but if they were born hundreds of miles apart this is much less likely. For nearly everyone today the distance between the places where they and their partner were born is greater than that separating their parents' birthplaces, and these parents in turn were almost certainly born further apart than *their* parents. In rural New England in the nineteenth century the distance between the birthplaces of marrying partners was less than 10 miles. Now the average is several hundred miles, and most American couples are completely unrelated.

It will take a long time before the mixing is complete—as much as five hundred years to even out the genetic differences between England and Scotland. Even if global homogeneity is a long way away, increased movement will certainly have a biological effect. No longer will large numbers of children be born who have two copies of defective genes because their parents share a common ancestor. There is little doubt that the most important event in recent human evolution was the invention of the bicycle. The bicycle means, too, that no longer is there much chance—as there once was in South Africa or Finland—for isolated populations to diverge by accident. One of the most important agents of evolution has gone forever.

All this suggests that social change—improvements in health and in mobility—mean that the biology of the future will not be very different from that of the past as mutation, natural selection, and random change lose their power. It may even be that humans are almost at the end of the evolutionary road, that we are as near to our biological Utopia as we are likely to get. Fortunately, no-one reading this article will be around to see if I was right.

J. S. JONES

Born 1944 in Aberystwyth and went to school in Liverpool. After obtaining his B.Sc. and Ph.D. at the University of Edinburgh, he went to the University of Chicago, and returned to London in 1971. He has worked on the ecological genetics of snails and slugs, and on behavioural genetics in *Drosophila*. At the same time he has maintained an interest in human genetics and evolution, and edited the *Cambridge Encyclopedia of Human Evolution*. He gave the 1991 Reith Lectures on *The Language of the Genes*, and a book with the same title was published this year.

Simple brains—simple minds?

MICHAEL O'SHEA

This essay arose from a public lecture at which I was able to demonstrate how living nerve cells work. No doubt the subject matter is difficult, but it was made accessible through the power of 'show-and-tell'. I hope here to be able to reproduce a semblance of the lecture and the exciting immediacy of that occasion. The central mission is to look at the brain as if it were a machine, to describe the components of the brain, and to explain how they work and interact. Importantly, we will consider how the 'simpler' nervous systems of invertebrate organisms can help us to understand the brains of higher animals, which are on the surface far more complicated.

The human brain is arguably the pre-eminent machine in the Universe. It is a machine with exquisitely complex and subtle properties. Figuring out how it works is not going to be easy. But this is the audacious objective that the brain, with characteristic immodesty, has set for itself. Of course the human intellect has a convincing track record in the biomedical sciences, providing more or less complete understanding of how the kidney, liver, and heart work, for example. But can the brain really expect to explain itself; can it understand the mind, consciousness, the sensation of free will, and so on? Or could it be that some problems may be inherently just too difficult and that the brain trying to understand the brain is like a person attempting to pull himself up by his own bootstraps?

On this question I prefer to be optimistic. Neuroscience will not be the first field of human endeavour to force us to grasp and manipulate counter-intuitive ideas. Consider, for example, modern astrophysics, a field in which experiment and theory have combined to generate concepts of space and time which have nothing to do with our everyday experience. Could our brains have evolved the capacity to conceive, grasp, and manipulate such concepts? What in our evolutionary past could have conferred a selective advantage on an animal capable of abstract leaps and bounds and the mental agility necessary to understand quantum mechanics, relativistic space–time, and black holes? The fact that some highly trained and talented minds can grapple with such concepts is reassuring because it indicates that ultimately the brain may indeed be capable of providing a complete explanation of its complicated self.

If the brain is a machine, constructed from components that obey the ordinary laws of physics, chemistry, and biology, then the extraordinary properties of the machine—mind, consciousness, and so on—must arise somehow from lawful interactions between the machine's components. From this point of view mental events, including all of human conscious experience, cannot be due to anything other than the interactions between definable components. Acceptance of this idea does not come naturally, even for a neuroscientist. Consciousness is a deeply puzzling fact of life. That it must be an emergent property of a machine does nothing to make it any the less puzzling. I console myself with the thought that water also has puzzling properties, but they all arise from nothing more than the interactions between oxygen and hydrogen atoms.

As the human brain is a machine, we ought to be able to learn something about how it works by examining the machine's components and observing how they interact. We should also be able to compare different types of brains in different organisms in order to gain insight into what components are required for the extraordinary feats of the human brain. For example, a comparison between simple and complex brains, made say between the brain of a worm or an insect and that of a man, might highlight features that subserve the higher mental capabilities characteristic of man but which are not a part of a worm's mental life.

There is nothing mysterious about mental events: they arise from the lawful interactions between the component parts of the brain machine. But what are the component parts, what do they look like, how do they work, what are the rules of interaction, and how are they organized to produce a biological computer? In brief, the components of the brain are highly specialized cells called neurones and in the human brain neurones are present in truly astronomical numbers. They are the components of intelligence, imagination, and consciousness and without them there simply is no mental activity. Neurones are such unusual cells that they are quite difficult to see using conventional microscopical techniques. For this reason, until the beginning of this century, the brain was thought not to be composed of cells at all. In organs other than the brain, cells are relatively easy to see with fairly straightforward techniques and not very sophisticated microscopes. Generally they are objects with regular, simple shapes and the idea that cells are the components of tissues is pretty self-evident under the microscope. But to the early microscopists the brain seemed to be an entirely different kind of organ, one without distinct cells.

I shall return to this shortly but now it may be instructive to consider not the microscopic components of the brain, but the brain's gross anatomy. We will examine two views of the whole brain, views separated by approximately 500 years. The first illustration is of a human brain drawn by Leonardo da Vinci (Fig. 1). While this is a beautiful drawing, it is not particularly accurate—in fact, it is vague and confused. Leonardo da Vinci was influenced by medieval philosophers who considered that there must be a tripartite division of the brain reflecting their idea of the division of mental functions into three aspects—imagination, judgement, and memory. In order to conform with this (wrong) idea, Leonardo draws three schematic ventricles as if these were the only components of the brain's machinery. These he labelled (in mirror writing incidentally) O, M, and N and he decided, by an unrevealed logic, that ventricle O was concerned with imagination, N with judgement, and M with memory. To say the least, this is not a particularly helpful theory of how the brain works—there appears to be no plausible mechanism. Today we know that mental activity has virtually nothing to do with what goes on inside the ventricles, but has a great deal to do with what goes on in the area of the brain in which Leonardo has indicated no particular structure at all.

Five hundred years on, it is possible to view the whole living brain by an imaging technique called magnetic resonance scanning. In the series of images in Fig. 2 it is possible to see what Leonardo da Vinci chose not to draw. I am happy to report that the inside of the brain (and the one illustrated happens to be my own) is not a void—the scan reveals intricate substructures. It is not the aim of this discourse to describe the anatomy of my brain or to say what mental functions might be associated with each of the major substructures. The point of showing both images is that, despite the far greater detail in the second, neither provides any clue as to how the brain machine works. In neither image can we see what could be regarded as the plausible components of mental activity. A machine is complex as the brain must have lots of different components and since the machine is quite small but very powerful, the components must be very numerous and very small—far too small to be seen without a microscope.

The term 'cell' implies uniformity, but the cellular components of the brain, the neurones, are very far from regular in shape; they have exceedingly fine and profusely branched processes which ramify from the cell's body and intermingle among the branches of other neurones. When a brain slice is viewed through an optical microscope using a conventional

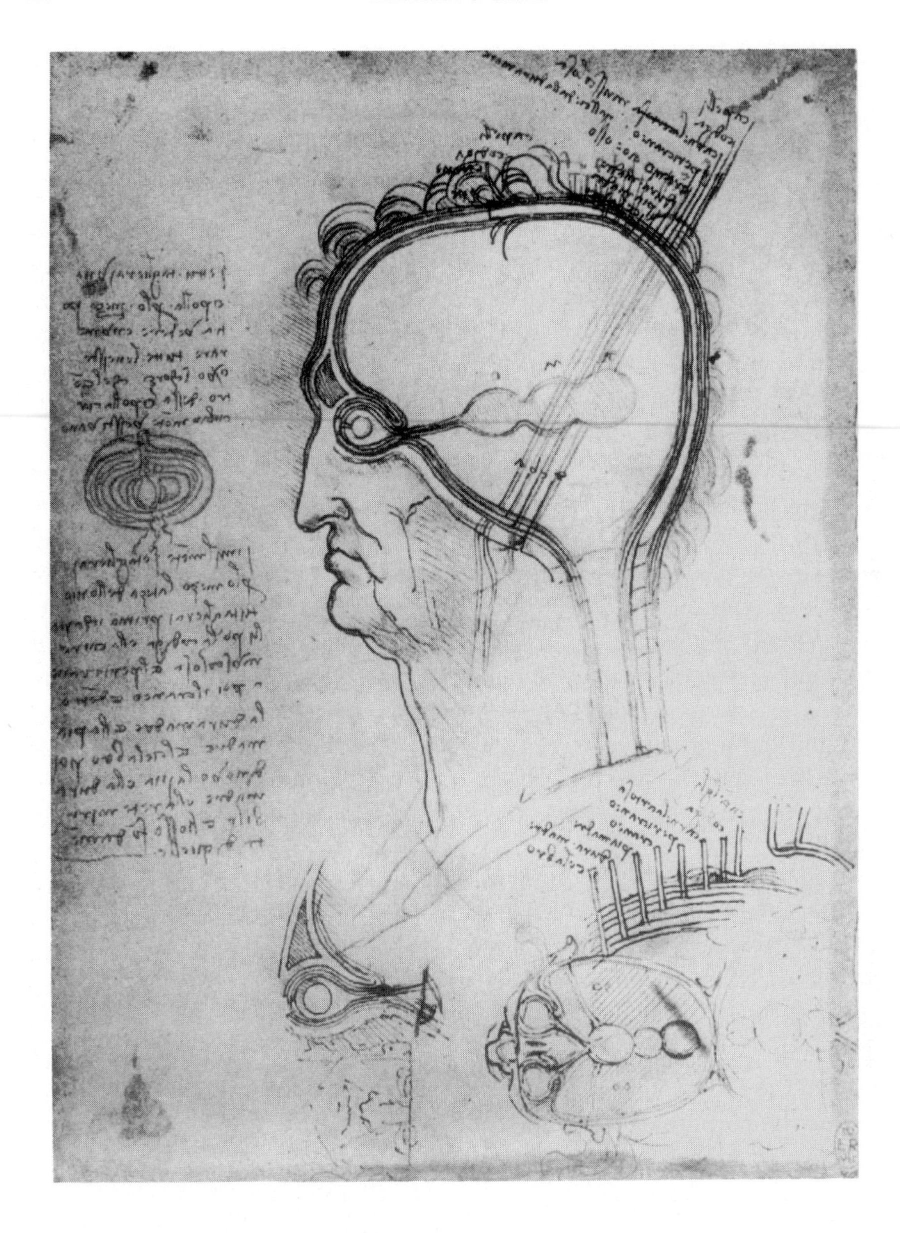

Fig. 1 A diagrammatic sagittal section of the head drawn by Leonardo da Vinci, probably around 1490. It represents an attempt to translate into drawings the description of the brain given by Avicenna. The smaller drawings on this page include a section of an onion, a section of the eye and orbit and a horizontal section of the head.

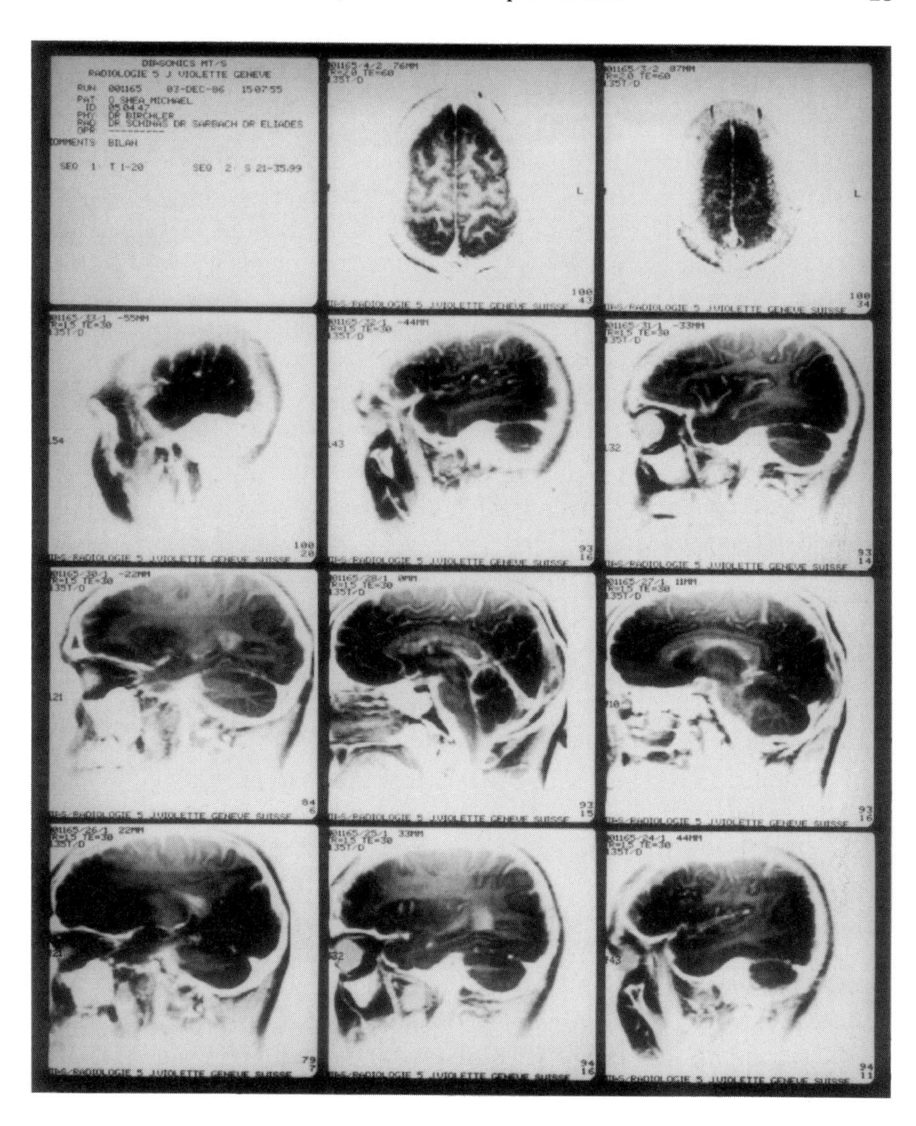

Fig. 2 A series of section images through the brain of the author. A near midline or sagittal section is shown in the middle of the third row. The folds of the cortex, the brainstem, and the cerebellum are clearly visible. These images are reconstructed from the nuclear resonance produced in a very strong pulsed magnetic field. I was surprised that the generation of these images produces no mental experience or sensation of any kind, except the regular thumping sound of the pulsed magnetic field.

staining method, one sees the nuclei of the cells' bodies embedded in what appears to be a hopelessly tangled morass with no evidence of the distinct cell boundaries that are so evident in other tissues. It is no wonder that the idea that tissues and organs are composed of cells was not thought to be applicable to the brain. The microscopic structure of the brain resembled a fine and non-cellular network and this image suggested a 'reticular' theory of brain anatomy that persisted well into the twentieth century.

For progress to be made, a staining method that would selectively highlight the entire structure of very few neurones in any particular region of the brain was required. Such a method would allow the microscopist to view a single neurone, unobstructed by the tangled mass of branched processes of neighbouring neurones. Eventually, a method to fit the bill was developed towards the end of the last century by the Italian anatomist Camillo Golgi. Golgi's method incorporates aspects of the chemistry of photographic processing and it reveals individual neurones as dark, silver-impregnated silhouettes, against an otherwise unstained background. Interestingly, even to this day no-one quite knows how or why Golgi's remarkable method works in such a highly selective manner. Nevertheless, for the first time the cellular components of the central nervous system were seen in their entirety, and what a remarkable revelation this must have been. Immediately it was apparent that neurones are discrete cells, but they are unlike any other cells in any other tissue. Most noticeably they differ markedly one from another, in particular with respect to the pattern of their numerous branched processes (Fig. 3, left side). Golgi's method revealed the machine's components and it provided the key to establishing a scientifically testable idea about how the brain works. It also signalled the end of the reticular theory and a far more powerful theory was born, namely that the brain, like other parts of the body, is composed of discrete cellular components.

It is perhaps difficult today for us to understand why the idea that the brain is composed of discrete cells should have been so difficult to con-

---

Fig. 3 At the top are drawings of a selection of neurones from the cortex of a young child. They have been stained using the method developed by Golgi at the turn of the century, but these drawings are by his contemporary Ramone y Cajal. At the bottom are drawings of a selection of neurones from the brain of an insect. They also were stained using the Golgi method and are the work of the contemporary neuroscientist, Nicholas J. Strausfeld. Notice that neurones differ markedly one from another, that there is no 'typical' neuronal shape, and that human and insect neurones appear to be at least equally complex.

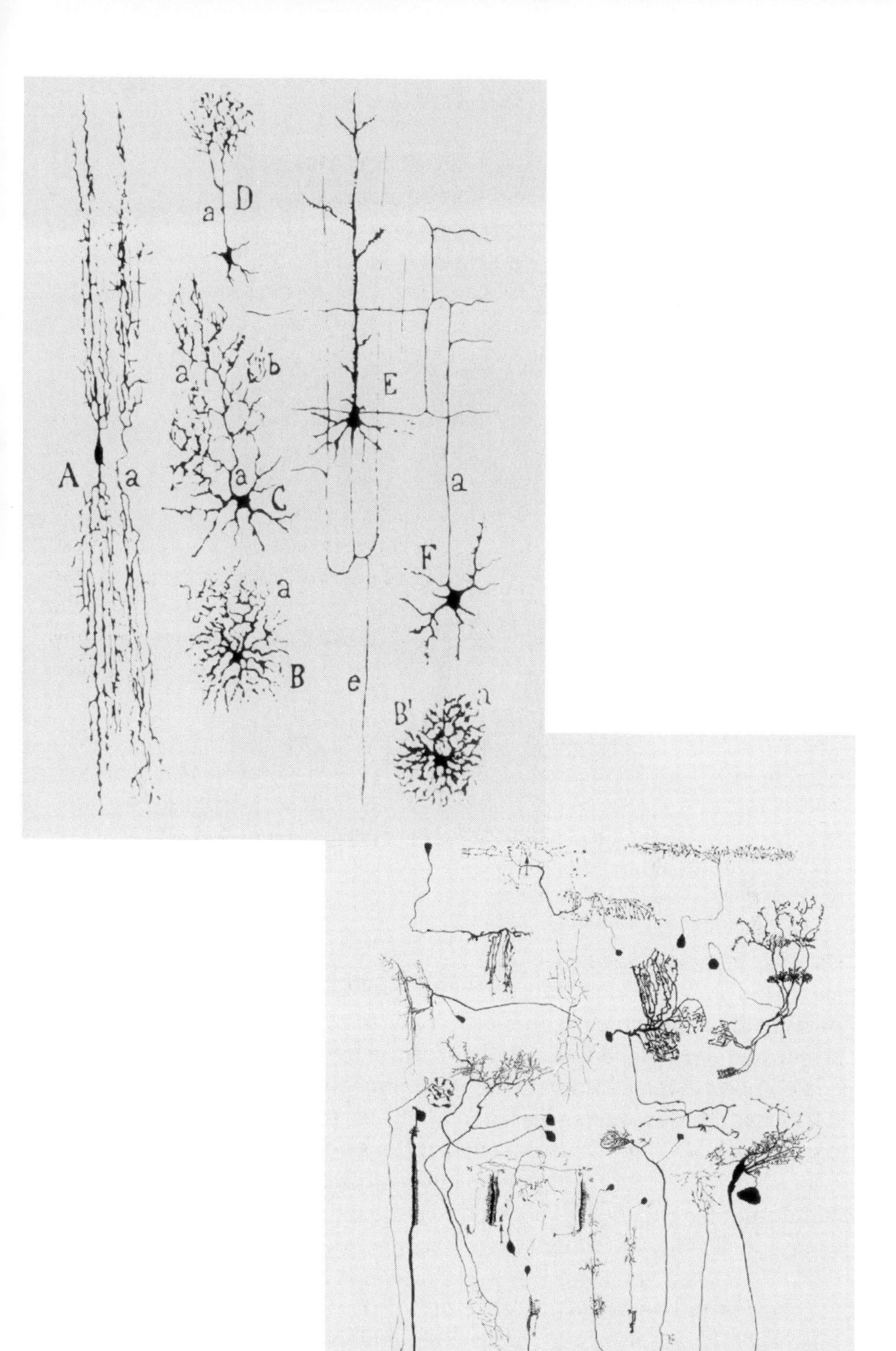

a
D
a
b
E
A
a
a
C
a
a
F
a
B
e
B'
a

Fig. 4 A photographic self-portrait of the great neuroanatomist Santiago Ramone y Cajal who applied the method developed by Golgi to the analysis of the brains of many different animals and man. He and Golgi shared the Nobel prize in physiology and medicine in 1906.

ceive and accept. Golgi himself, in spite of the images provided by his own method, remained convinced of the reticular theory of the microscopic anatomy of the brain. Rather surprisingly, it was not Golgi but his contemporary, the Spaniard Santiago Ramone y Cajal (Fig. 4), who made the most profound inferences based on his observations of the brain using the method of Golgi. Cajal established the so-called Neurone Doctrine—a set of ideas that for the first time allowed us to think about the brain as an information processing machine constructed from discrete microscopic components whose function was to transmit, transform, and store information.

Cajal not only studied the anatomy of the human brain, he was also interested in comparing the microscopic anatomy of the brains of markedly different organisms. In particular he was interested in the brains of insects. There is of course an enormous gulf between human and insect intelligence, so a comparison of their brains might suggest what structural features of the brain's components perform the complex information processing functions that are uniquely human. One might reasonably expect, for example, the human brain to be constructed from 'high performance' components and the insect brain to contain markedly less sophisticated components. A selection of neurones in an insect brain, as revealed by the Golgi method, is also shown in Fig. 3 (right side). These are modern drawings by the British neuroanatomist Nicholas J. Strausfeld, but Cajal saw similar images and was struck by the variety and complexity of the forms of the insect neurones. Certainly the components do not appear to betray the gulf between insect and human intelligence—the insect neurones are equally complex, they display as much diversity of form, and they are about the same size. Cajal himself expresses considerable surprise at this, as the following quote indicates. He said, '. . . the quality of the psychic machine does not increase with the zoological hierarchy. It is as if we are attempting to equate the qualities of a great wall clock with those of a miniature watch.'

Our brains are conscious machines, but the components apparently are not special. This is puzzling, as Cajal himself noted, because if the components are equivalent in such different organisms, on what can the very special properties of the human brain depend? To answer this question, we need to know how neurones work and interact, and there is still a great deal to discover before we will understand these fundamental processes that underlie mental activity. Fortunately, we do not need to study this aspect of brain function in the human brain. Simpler organisms such as worms, molluscs, and insects have provided and continue to provide us with model systems in which to study the cellular and molecular mechanisms fundamental to the operation of all brains, including our own.

Neurones appeared on this planet in the brains of simple organisms hundreds of millions of years ago, long before the evolution of intelligent, conscious animals such as ourselves. With the appearance of neurones, evolution was set on a remarkable course, because when present in very large numbers, assemblages of interacting neurones become a conscious, self-reflecting machine. Like it or not, the astonishing conclusion from comparative studies is that the evolution of a biological machine capable of the extraordinary feats of our own brains did not require the evolution of fundamentally new types of component. Contrast this with the dramatic improvement in the performance of computers that has explicitly depended on the invention of entirely new components. Early computers and calculating machines, for example, were mechanical. More recently a series of electrical inventions (vacuum tubes or valves, the transistor, the microprocessor) have heralded the introduction of new generations of machines, each dramatically faster and more powerful than its predecessor.

In the hands of engineers, computers developed additional power by design. If computers had been developed as brains have evolved, they would still be made of mechanical cogs and levers!

The neurone was a fantastically successful 'invention'. In its fully developed form it is several hundred million years old, but intelligence and consciousness did not appear on this planet before one million years ago. Clearly the neurone was not 'invented' for intelligence or consciousness, but both could arise through neuronal interactions in complex brains without the need for any other special component. What appears to underlie the superior performance of complex versus simple brains has a great deal to do with the sheer number of components and little or nothing to do with their type. Consider, for example, the simple nervous system of a nematode worm (a roundworm). It contains only a few hundred neurones, whereas the human brain contains 100 billion—about the same as the number of stars in our galaxy. Truly astronomical. Simpler brains can help us to understand how the human brain works because their components are equally complex but they occur in far, far fewer numbers, allowing us to study how neurones interact with one another and how they perform the fundamental tasks essential to mental activity. What a worm's brain does in a small way, ours does in a big way!

The function of neurones is to transmit, transform, and store information and there are two interacting processes by which this is achieved—one is principally electrical and the other is chemical. Information is transmitted along the branched extensions arising from the neurone's cell body (known as the soma) in the form of brief electrical disturbances or pulses. The neurone's branches may be thought of as electrical cables or wires, so let us explore this analogy a little further in order to see what problems had to be overcome by the evolutionary process to produce a biological device capable of transmitting electrical signals along microscopically fine cables. We will examine the neurone's axon, the name given to the longest extension which carries electrical signals away from the cell body. Axons of some neurones can be very long indeed. For a neurone located in the spinal cord of a whale or giraffe that is connected to skeletal muscle (a motor neurone), the axon would have to be many metres in length. How can electrical signals be transmitted reliably by a cell over such distances?

The axon resembles an electrical cable because it has an electrically conductive core enclosed by an insulating layer formed by the cell membrane which is composed of rather non-conductive fatty or lipid molecules. So can the neurone simply initiate an electrical signal at one end of the axon to have it appear at the other end essentially unaltered, with no significant distortion or loss of strength? Unfortunately, although insulated electrical wires can be relied upon faithfully to transmit electrical pulses over very considerable distances, axons cannot. There are a number of problems. First, the core of an axon is not a particularly good electrical conductor, being millions of times less conductive than copper wire. Secondly, the axon is not particularly well insulated and, to make matters worse, it is bathed in a highly conductive salty solution, the extracellular fluid, that resembles sea water in its salt content. All of this means the electrical signal passing along that axon is progressively diminished and distorted by the combined effects of the inadequate insulation and the high capacity of the axon's membrane to store electrical charge, slowing the rate at which the voltage signals can be generated. As a signalling device the axon would appear to be completely inadequate for the job, as it is a lousy electrical cable! Just how bad it is can be judged by a direct comparison with real electrical cables. If you initiate an electrical pulse in a length of 22 gauge copper wire, it can be detected in the wire 10 billion miles away, at which distance it will have suffered only a 66 per cent loss of its original strength.* In an axon the pulse would have diminished in strength by 66 per cent in less than 1 mm, and another 66 per cent loss in the next millimetre. A rather disgusting state of affairs for an axon trying to communicate with even a nearby target, let alone with the tail fin of a blue whale several metres away!

So we seem to have evolved a biological device for electrical signalling with appallingly bad electrical properties. The comparison with the length of 22 gauge copper wire, stretched across the vacuum of space, however, was a little unfair on the axon. A much closer analogy can be drawn between an axon and a submarine telegraph cable—both are insulated conductors surrounded by a highly conductive salty medium. The copper core of the submarine cable is very well insulated from the sea water, but the insulation is not perfect and so some of the signal inevitably 'leaks' out and is lost in an ocean of electrically very conductive salt water. The rate of signal loss in submarine cables is a serious problem because, while the transmission distances are counted in thousands of miles, about 66 per cent of the original signal strength is lost every ten miles or so. To solve this problem, booster stations are designed into submarine cables, spaced so that the failing signal can be detected, reshaped, amplified, and sent on its way to the next station. Some hundreds of millions of years ago an almost exactly analogous solution to the

* In order to compare the performance of different cables, a parameter called the 'length constant', or λ, has been defined. One length constant is the distance over which the amplitude of an electrical signal diminishes to $1/e$ of its original strength. This represents a loss of about two-thirds and this is why I have used 66 per cent loss as a way to compare electrical and biological cables.

problem of electrical signalling along axons was found by the purposeless process of evolution.

The mechanism of transmitting electrical signals along axons is today almost completely understood. The most significant contribution in this area was made by the two great British physiologists A. L. Hodgkin and A. F. Huxley in the late 1940s. Hodgkin and Huxley studied an axon in the nervous system of a marine mollusc, the squid. This animal was selected because an anatomical investigation by the British zoologist J. Z. Young identified a giant axon running more or less the entire length of the animal's mantle. Axons are normally exceedingly fine structures, often less than one 10 millionth of a metre in diameter, making physiological study very difficult. The axon described by Young, however, was about 1 mm in diameter, so surprisingly large, in fact, that Young at first mistook it for a blood vessel!

A series of famous and elegant experiments on the squid's giant axon followed, culminating in a theory explaining how the brief pulses of electricity used by neurones as signals, the so-called 'action potentials', are generated and propagated by the axon. This account of the action potential in the squid giant axon was to earn the Physiology and Medicine Nobel prize for Hodgkin and Huxley in 1956, not least because the principles and mechanisms uncovered were universal. That is to say with very minor exceptions of detail all axons, from lowly invertebrate organisms to man, operate in fundamentally the same way. Neurones apparently not only resemble one another in widely different organisms, they also share the same operating principles.

It is not possible here to explain fully how action potentials are produced but I will try to give you a flavour of this process. We have seen that the submarine cable can transmit signals over great distances by providing power to the cable at intervals to amplify the decaying signal and send it on its way. Similarly, the axon must be provided with a source of power so that signals can be propagated along the axon without loss of strength or distortion, irrespective of the length of the axon. Power is derived from the capacity of cells, including neurones, to couple the expenditure of metabolic energy to the movement of electrically charged particles (ions) across the cell membrane. This is achieved by an enzyme which resides in the neurone's membrane and 'pumps' sodium ions (Na^+) out of the cell and potassium ions (K^+) into the cell. The net effect of this is to produce two concentration gradients of positively charged particles, creating a situation in which (to rebalance the concentration gradients) positive sodium ions tend to move into the cell and potassium out. Essentially we have created two biological batteries spanning the neurone's membrane, one with its positive pole facing inwards (the sodium battery) and the other with its positive pole facing outwards (the potassium battery). The actual voltage across the membrane at any time is determined by the degree to which each battery is activated, and this is determined by the relative flow of potassium and sodium across the membrane. The two transmembrane batteries are used to power an amplifier of molecular proportions in the axonal cable, an amplifier which detects small electrical signals and then boosts them as they pass along the axon. To understand how this is done we need to know the values (in volts) of the batteries. Also we need to know how some of the proteins that span the fatty membrane enclosing the axon change their properties in response to changes in transmembrane voltage.

The sodium battery is about +0.05 volts and the potassium battery is about −0.07 volts (opposite polarities, voltages referenced to the inside of the axon). As mentioned above, by altering the relative flow of sodium and potassium the actual voltage across the membrane can fluctuate and you can see now that it can do so only between −0.07 and +0.05 volts. When the neurone is not generating a signal, the membrane is said to be at rest and then the membrane potential is about −0.07 volts; this means the potassium battery must be far more active than the sodium battery. When a neurone is stimulated or excited, the membrane potential is caused to be less negative. This shift of voltage leads to the activation of sodium flow into the neurone and to a further shift of voltage towards the potential of the sodium battery and more activation of sodium flow. The more the potential shifts towards the sodium battery potential, the more the flow of sodium is activated and the inside of the axon rapidly becomes positive. Almost as rapidly, however, two other mechanisms are set in motion, one leading to the inactivation of sodium flow and the other to the greatly increased activation of the flow of potassium. Both combine to cause the membrane voltage to return to the resting level just before it reaches the value of the sodium battery potential. These events are first initiated by a small change in the membrane voltage which is then amplified by ion flow into a brief electrical pulse.

Today the molecular machinery underlying signal detection, and amplification and the generation of the propagated action potential is well understood. Sodium and potassium flow into and out of the axon through separate pores or channels in the membrane formed by protein molecules which span the membrane. Briefly, the protein channels can be either closed or open, thus either preventing or allowing ion flow. Importantly, the proteins are endowed with a special property called 'voltage sensitivity' which means that they can detect small changes in the transmembrane voltage and respond by opening or closing. The sodium channels are mostly closed at the resting potential but in response to small shifts of the voltage (less negative), they open. The resulting sodium inflow leads to further voltage change and to the opening of more and more sodium channels. Similar changes occur to the potassium channels which also tend to be closed at the resting potential. The potassium channels, however, open more sluggishly so that the sodium inflow precedes potassium outflow. Thus when excited by a small electrical disturbance the membrane moves from the resting potential, towards the value of the sodium battery (+0.05 volts) and then returns to the potassium battery potential (−0.07 volts). For biological signalling systems these changes are extremely rapid; the action potential lasts just a few thousandths of a second and can travel along the axon at up to 100 metres per second.

As we have seen, the key to understanding the molecular machinery behind the amplified action potential is to be found in the voltage-sensitive proteins that form channels in the neurone's membrane. These proteins would appear to be very sensitive indeed to changes in voltage; a change of just a few thousandths of a volt across the membrane can trigger the opening of a gate of molecular proportions, leading to more current flow and amplification into a full action potential. How can the protein channels function as voltage sensors able to detect small changes in voltage and to amplify them into action potentials? Voltage sensitivity is achieved by the presence in the channel proteins of electrically charged amino acids. When the voltage field across the membrane changes, the charged amino acids are forced to move and their movement alters the three-dimensional geometry of the protein, causing it to adopt the 'open' configuration. But surely these dramatic changes in the behaviour of a protein could not be triggered by voltage disturbances of just a few thousandths of a volt. Actually it is a mistake to think that the electrical forces acting on the channel proteins are weak or feeble, as the following calculations will show. The total voltage change during an action potential is about ±0.1 volt (from about −0.07 to +0.05 volts and back). This relatively small voltage fluctuation, however, occurs across a membrane a mere one millionth of a centimetre thick. During the action potential, therefore, the channel proteins experience the reversal of an enormous voltage gradient of 100 000 volts per centimetre in approximately one thousandth of a second! From this perspective it is somewhat less remarkable that the channel proteins respond dramatically to small changes in the absolute transmembrane voltage.

Information flow along axons, as we have seen, is electrical but communication between neurones is not. One of the natural consequences of the Neurone Doctrine (that neurones are cells) is that neurones cannot be joined together directly, and yet they must communicate with one another. This is achieved by the release of chemicals called neurotransmitters and this occurs at specialized structures called synapses. Most types of neurotransmitters are stored and packaged ready for release in tiny vesicles concentrated on the transmitting side of synapses. On the arrival of an action potential at the synapse, the vesicle membrane fuses with the neurone's membrane and the transmitter contents spill into the gap or 'cleft' between the transmitting (presynaptic neurone) and receiving (postsynaptic) neurone (Fig. 5). By this process electrical activity in the transmitting neurone is coupled to the reception of a chemical signal in another neurone.

At the first level of complexity, neurotransmitters are either excitatory or inhibitory according to whether they initiate electrical activity (excitatory) in the receiving cell or reduce the probability of electrical activity (inhibitory). If this were the whole story, there would be little justification for there to exist more than two neurotransmitters—one excitor and one inhibitor ought to be sufficient. In fact things are a lot more complicated than this; transmitters cannot be classified so simply and there are hundreds of different types of transmitters grouped in distinctly different

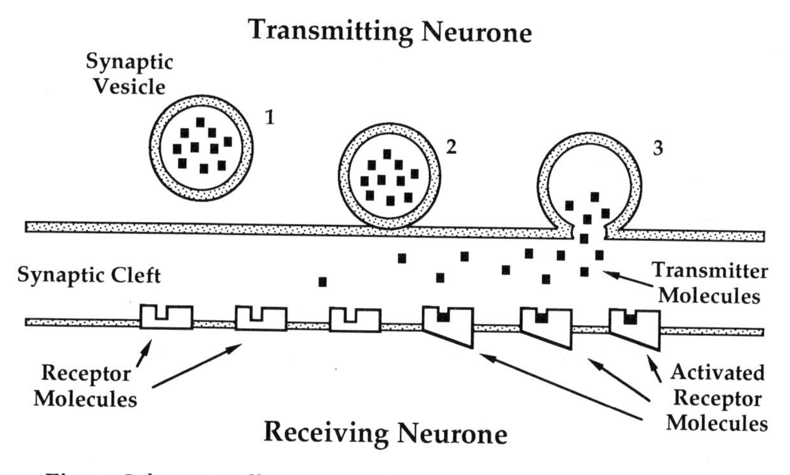

Fig. 5 Schematic illustration of a synapse. Vesicles located in the transmitting neurone, on the presynaptic side of the cleft, are caused to be mobilized (steps 1 and 2) on the arrival of an action potential. The membrane of the vesicle then fuses with the neuronal membrane and the neurotransmitter molecules spill out into the cleft (step 3) where they interact with receptor molecules causing them to be activated. This illustration is not drawn to scale, but the vesicles range in size from 40 to 200 billionths of a metre in diameter. The receptor and transmitter molecules have been drawn many times larger than they would appear relative to the vesicles. Receptor activation which follows transmitter binding probably involves a change in the shape or conformation of the protein.

chemical families. Neurotransmitters can be relatively small organic molecules such as acetylcholine, or amino acids and molecules derived from amino acids such as dopamine, adrenalin, and serotonin. The largest single class of neurotransmitter substance are fragments of proteins called peptides. It is estimated that there are one hundred or more different neuropeptides in the human brain, and a similar number in the brains of far simpler organisms such as snails and insects. Each plays a different role, and we shall return to discuss what they may do later.

The chemical message is interpreted in the receiving neurone by highly specialized protein molecules, sensibly named receptors. We can understand what receptors do quite easily. Their first task is neurotransmitter recognition. Receptors are not promiscuous: generally each receptor is 'tuned' to react to only one neurotransmitter—a highly specific process often compared to lock (receptor) and key (neurotransmitter) mechanisms. Recognition is followed by receptor activation and a response in the receiving neurone. Receptor activation is terminated when the transmitter is removed from the synaptic cleft either by being destroyed by enzymes or by being absorbed by nearby cells.

As indicated above, the simplest type of response to a neurotransmitter is electrical excitation or inhibition. So the most basic mode of information transmission in the brain is from electrical to chemical to electrical, from neurone to neurone. But how is a chemical signal transformed into electrical excitation or inhibition in the receiving neurone? There is a clue to the solution of this problem in the explanation of the action potential. Here voltage disturbances are amplified by the flow of ions through voltage-sensitive channels in the membrane. If the activated receptor protein were able to form a channel in the membrane through which positive or negative charge (in the form of ions) could flow, then we would have a very direct way to link neurotransmitter action with the electrical excitation or inhibition. In fact, members of a major class of receptors operate in precisely this manner—they have been called 'transmitter-gated'—and such receptors directly mediate the electrical to chemical to electrical flow of information in the brain. Unfortunately, this does not account for all information flow in the brain as many neurotransmitters do not simply cause excitation or inhibition and they do not gate ion channels. Most neurotransmitters carry information of a far more subtle type and do so by causing the synthesis of other chemical messengers in the receiving neurone. Such chemicals are called 'second messengers', to distinguish them from the primary messengers that pass between neurones, the neurotransmitters.

Importantly, the second messengers provide for change and plasticity in the interactions between neurones, and it is becoming clear that any understanding of adaptive behaviour will depend on understanding the subtle roles played by them. Their discovery means that the brain is most certainly not reducible to a fixed wiring diagram in which there is simple electrical to chemical to electrical flow of information between neurones. Interactions between neurones certainly do not simply involve the transient increase or decrease in the probability of electrical activity. Clearly, an understanding of the subtle roles played by transmitters that regulate second messenger synthesis will provide the key to gaining a deeper understanding of how the brain machine works.

Second messenger synthesis is capable of activating a cascade of biochemical events in the neurone that far outlasts the electrical consequences of the 'simple' excitatory and inhibitory type of neurotransmitters. Second messengers can interact with the electrical properties of neurones, causing for example long-term changes in the excitatory or inhibitory potency of the 'simple' transmitters and in this way pathways in the brain can be facilitated or suppressed. The streaming of information in the brain therefore is constantly being modified and shaped by previous activity and a fixed 'wiring diagram' of the brain does not exist in any simple sense, as it can for a computer. It is my guess that for computers to approach the power of the brain, their designers must find some way to emulate the complexity and subtlety that the modulatory neurotransmitters and second messengers overlay on the brain's basic wiring diagram.

Viewed in this way the brain is an electrical machine modified and modulated by chemistry. This electrochemical brain-machine is of almost fantastic complexity and cannot be diagrammed or defined because it is constantly being shaped and reshaped by the action of the modulatory neurotransmitters acting through second messengers. What hope do we have of ever coming to grips with such an apparently mercurial machine? The fact is that we are still very far from understanding how neurones are organized to produce even the simplest of behaviours. Perhaps the simpler brains of invertebrate organisms can help us to understand how networks of neurones interact to coordinate simple behavioural acts as a first step in a broader understanding of how the brain machine is organized. Studies of invertebrates, as we have seen, have already provided the fundamental and universal laws defining how the individual components of the brain work. We now need to know the rules that govern the interactions between neurones in functional networks. If there are universal rules then in simple brains we may learn about the fundamentals that govern the operation of far more complex interacting networks of neurones.

Let us now look at neurones in a simple brain interacting in a very simple network to generate an interesting and potentially useful rhythmic pattern of activity. All animals must be able to produce simple rhythmic movements and such movements must originate somehow from the interactions between neurones in the nervous system (Figs 6 and 7). Surely it should be possible in the case of simply repeated rhythmic movements to show how neural networks are organized in their generation. The circuit (which can be found in a snail brain) illustrated in Fig. 7 shows how rhythmic activity can be achieved using only three neurones, but even this very simple circuit has some instructive features. Firstly, notice that one of the neurones seems to be endowed with the ability to generate an intrinsic rhythmic pattern consisting of periodic bursts of action potentials, without the need of any interaction between other neurones. So while the analysis of neural networks is clearly important, some patterns of output arise from the special individual properties of the individual components and are not a property of the neuronal network. Only in numerically simple brains can we hope to understand how networks work by taking into account, as this simple example demonstrates we must, the individuality of each component.

Another interesting feature of the three-neurone circuit is that one neurone is able simultaneously to excite and inhibit the receiving neurones. This can arise in one of two ways: either the transmitting neurone releases different neurotransmitters at different synapses or the neurotransmitter released is the same but the two receiving neurones possess different receptors. It is an unfortunate fact of life that in order to analyse even a simple neural circuit such as this we need to know more about it than the anatomical connectivity. As this simple example shows, we must know also about the special properties of the individual components and details of the types of response generated at each synapse. This is a daunting task for circuits of hundreds, thousands, or millions of components and not one that can in reality be achieved, except in very particular circumstances.

Impossible though a complete understanding of how exactly a circuit

---

Fig. 6 An individually identified neurone in the brain of a snail. The cell's body or soma is about 0.02 mm in diameter and can be seen in the upper right-hand lobe of the brain. It gives rise to a major process which turns to enter the lower right lobe where it divides to form axons leaving the central nervous system to innervate muscles. This is one of about 20 000 neurones in the snail brain, many of which have been uniquely identified and named. Circuits of interacting identified neurones have been analysed and the neural circuits required for some simple behaviours are now understood in considerable detail. Neurones like the one shown here participate in the simple circuit shown in Fig. 7. Image kindly provided by Paul Benjamin.

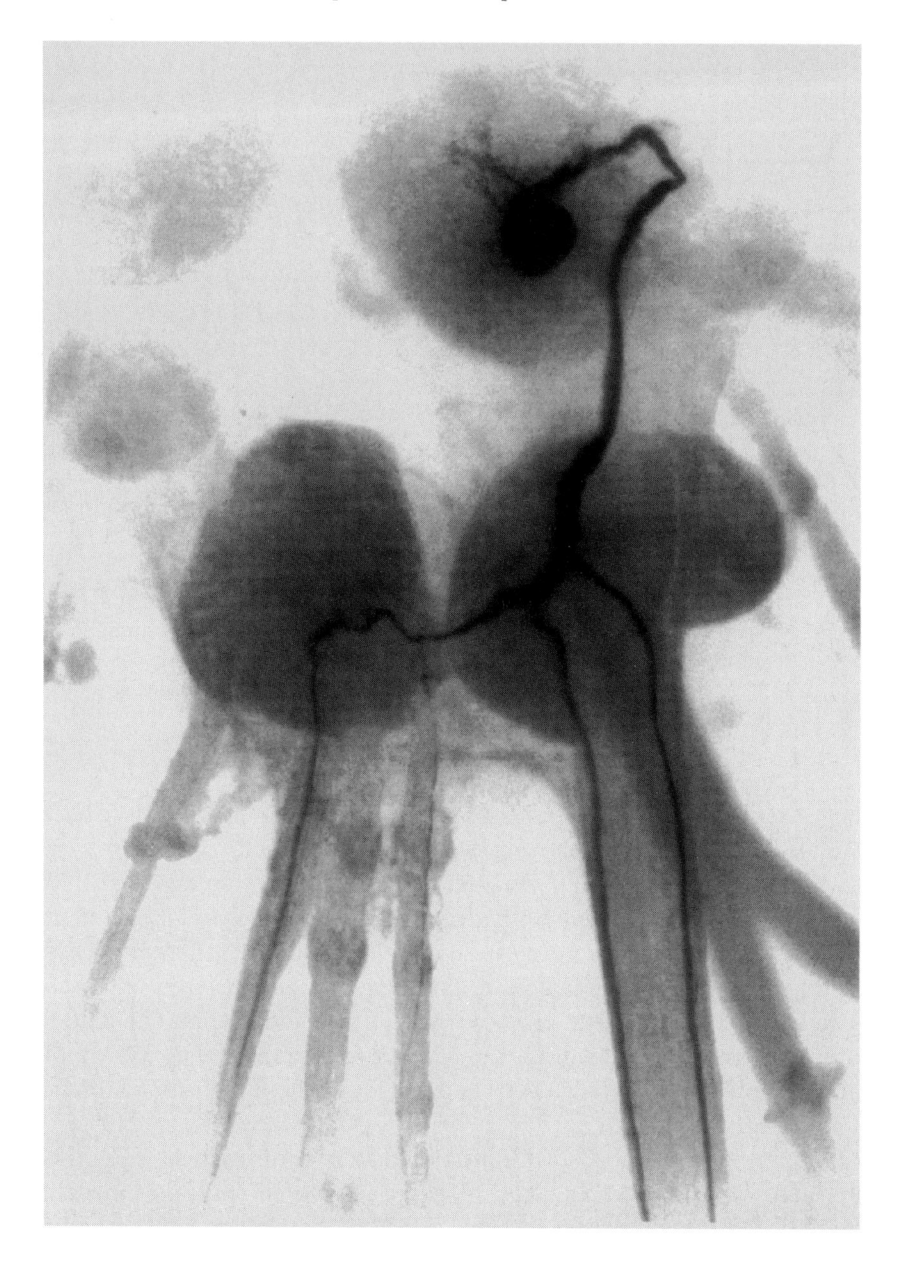

of neurones generates behaviour may now seem to be in practice, this is precisely what some neuroscientists are attempting. To do so they are using simpler invertebrate nervous systems. Importantly, recent work on 'simple' nervous systems has provided fascinating insights into what may be a crucial role for the many modulatory-type neurotransmitters. As we

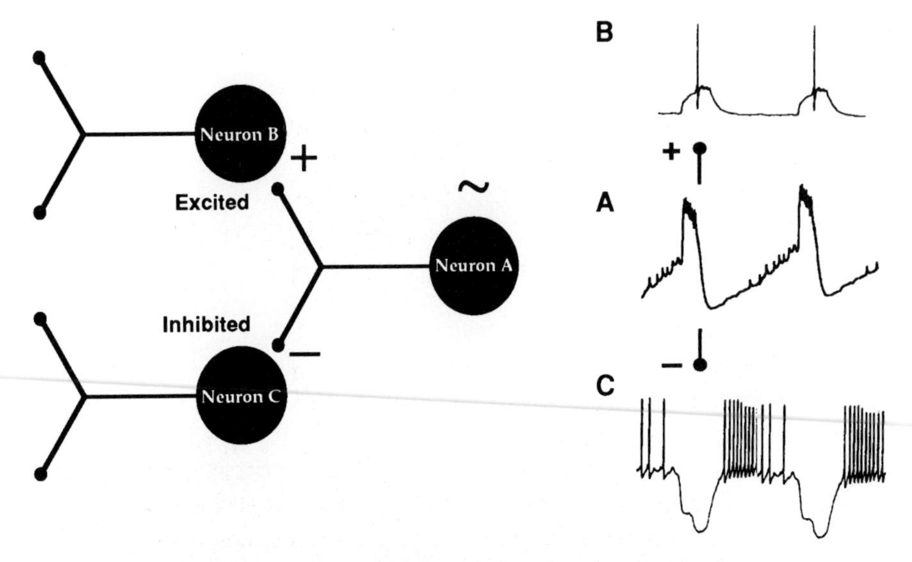

Fig. 7 Diagram of an actual neural circuit in the snail brain. Neurone A generates rhythmic bursts of action potentials and excites neurone B but inhibits neurone C. Simultaneous electrical recording from each neurone (on right) shows that neurones B and C are active in alternating bursts of action potentials.

have seen, such transmitters predominate but do not directly produce excitation or inhibition and are not therefore directly involved in the electrical wiring diagram or the anatomically defined neural circuit. The idea which seems to be emerging from detailed studies of simple neural circuits is that the modulatory transmitters are able to interact with the circuit, to sculpt and reconfigure it to generate either subtly altered or entirely different patterns of activity. It is becoming apparent that the output from a neural network is not fixed or predetermined by its synaptic connectivity and knowledge of the circuit alone therefore is not sufficient to predict its output.

Among the preparations which have led the way in developing these new ideas is a network of only 30 neurones which controls the movements of the foregut of crustaceans. All 30 neurones are contained in a small but accessible ganglion which can be isolated from the rest of the nervous system and for this reason it has been the subject of intense investigation for more than 20 years. Now it is more or less completely defined, all 30 neuronal components are individually identified and their pattern of synaptic interconnectivity is known. The anatomical wiring diagram of this circuit is of course, far more complicated than our three-neurone circuit illustrated in Fig. 7 and it is not surprising, therefore, that it is capable of producing more than one pattern of activity. When isolated from all other influences* the 30-neurone network generates two basic rhythmic patterns, one called the gastric mill rhythm and the other the pyloric rhythm. It seemed that the two patterns of activity could be simply explained by the existence among the 30 neurones of two semi-independent wiring diagrams. The conclusion seemed to be that different circuits are used for different behaviours. If this were indeed all that there was to it, one would have to say that such a result is hardly likely to change the way we think about how the brain works.

As no doubt you may have guessed, however, the simple output patterns of the isolated ganglion of 30 neurones conceal a far more complicated and interesting picture. When the activity of the ganglion was investigated in the intact animal, the gastric mill and pyloric rhythms were seen to be highly variable, with changes of timing, phasing, and intensity occurring during and following feeding. Remember that the network consists of only 30 neurones but now we must accept that it can produce several, not just two, different patterns of activity. Conceivably this could be due to the presence among the components of the network of several internal circuits that can be activated independently. This is unlikely to be the case.

The 30 neurones of the anatomically defined circuit are in fact the target of some 60–120 modulatory neurones that are not themselves part of the network but are located in different parts of the nervous system. While the 30 'circuit' neurones operate using only two neurotransmitters, the modulatory neurones use some 20 different neurotransmitters, many of them neuropeptides. Different patterns of activity arise from the 30-neurone network because different modulatory neurotransmitters modify the individual physiological properties of the circuit neurones and also alter the strength of their synaptic connections. So while not altering the anatomically defined circuit, the modulatory neurones can effect important changes in the components and in the synapses, changes which amount to the 'sculpting' of a variety of different functional circuits from a single anatomically defined network. It has been necessary for me greatly to simplify the picture which is now emerging from these studies, but you can perhaps see that it may be necessary to distinguish between two fundamentally different neuronal entities—the modulated and the modulators. In the example I have been describing, which is arguably the most comprehensively studied of all neural networks, it has been helpful to consider the modulated neurones and the modulator neurones

* This group of neurones can be studied in isolation because it is in a ganglion that can be dissected from the animal and maintained in a viable state while microelectrodes are inserted into selected identified neurones.

as separate and independent. For the moment this is a convenient simplification which helps us to begin thinking, in a new and hopefully fruitful way, about the brain and how it works. From a number of studies of the simpler brains of simpler organisms we are beginning to formulate an impression of great flexibility in the organization of neural circuits. As we have seen, the 30 neurones of the crustacean ganglion, for example, can be functionally reconfigured to produce several quite different patterns of activity by modulatory neurones that outnumber the modulated by about 4 to 1. It remains to be seen whether such studies on numerically simple nervous systems will 'scale up' and help us to understand the rules of engagement when the numbers of components are counted in billions. The fact that the individual components and the chemical messengers in simple and complex brains do not differ significantly, encourages me to believe that the rules of engagement may not be different either. Indeed, if they are significantly different we will have no simple models and I wonder how else we might gain a deeper understanding of how the hypercomplex human brain machine works.

We have already learned a great deal about how the components of the brain work by studying simple brains. But the most puzzling properties of our brains, namely mind, consciousness, and the like, emerge from interaction with complex networks of components and between the brain and the outside world through our emotions. We are just beginning to understand how neural networks are organized in simple organisms, and principles of organization are now being formulated. If many functional circuits can be sculpted from neural networks in simple brains, how much more powerful must these machines be than the small number of components would suggest. All organisms, even the most simple, have nervous systems that provide for adaptive behaviour. We know now that even 'simple' interconnected networks of neurones are not 'hard wired' like man-made electronic machines and that adaptive behaviour in simple organisms emerges from the continuous reshaping, reconfiguring and sculpting of circuits. Perhaps some of the more complex emergent properties of our own brains also appear by similar processes occurring in numerically far more complex circuits. Ultimately we must formulate rules that govern how large interacting neural networks work. I have no doubt that future studies on simpler brains will have a critical role to play, just as they have already had a critical role in defining how the individual components work.

Acknowledgements

I thank Dr Mark Yeoman, Dr Matt Brierley and Dr Richard Rayne for their invaluable practical assistance during the Discourse. They and others, most notably Professor Paul Benjamin, Dr Maurice Elphick, and Dr Jenny Rusted, have advised me on the presentation of the material included in this essay and have helped me to find ways to explain difficult concepts in straightforward language. I thank Annie Bacon for preparing the manuscript. The Sussex Centre for Neuroscience is supported by a grant from the Biotechnology and Biology Science Research Council.

MICHAEL O'SHEA

Born 1947, educated at Forest Hill School, London, and the University of Leicester where he received a first in Biological Sciences (B.Sc. 1968). From 1968–71 he was at the University of Southampton, where he obtained his Ph.D. (1971). From 1971–1974 he was a NATO post-doctoral fellow at the University of California at Berkeley, and from 1974–1976 a post-doctoral fellow at the University of Cambridge. In 1976 he was Assistant Professor of Neurobiology in the School of Medicine at the University of Chicago. In 1984 he was nominated Professor of Neurobiology at the University of Geneva, Switzerland and in 1988 returned to the UK as University of London Professor of Molecular Cell Biology. In 1991 an Interdisciplinary Research Centre in Neuroscience was established at the University of Sussex and he was appointed as the Founding Director. In 1990 his only child Linda, aged 11, died from an incurable brain tumour, strengthening his resolve to promote our understanding of basic brain mechanisms. He would wish, were there more hours in the day, to devote time to cycling, running, mountaineering, maintaining an ancient Lotus Elan, and listening to Mozart. He is a Life Counsellor of the NSPCC and was elected Fellow of the Royal Society of Arts in 1992.

Dead or alive: two concepts of rock behaviour

TIM HARPER

Living things such as ourselves develop concepts of the processes and objects we encounter. The adequacy of these concepts is critical to our individual and collective progress, even survival. In this article I will try to convince you that the same entity or process can be viewed in radically different ways.

I shall use rock as the entity or process of which we develop the concepts. I shall examine two opposing concepts of the nature of rocks, and shall use the engineering of rocks as a way of measuring the utility of these concepts.

On the one hand we can adopt the common perception of rocks as passive, clod-like, inert, Newtonian entities. This traditional perception can be helpful to us, and I use a process called hydraulic fracturing to illustrate this. I then examine rock behaviour more closely in the laboratory, and show that it is necessary to identify a new concept of rock behaviour which is more compatible with our observations. This new concept pictures rock as a system of vast numbers of interacting elements. It is the interaction of these elements which governs the system behaviour, not the substance of the particles, nor, from some points of view, the details of the rules governing the interaction of the individual sand grains. Rock responds in a similar manner in response to many different forms of disturbance—it evolves to a state in which it is remarkably sensitive, ready to facilitate the next flux of energy or matter through the system.

It is critical that the multiplicity of interactions occur not just internally but also *externally*, with the environment outside the boundaries of the system. Let us not forget that these boundaries are nominated solely for our convenience. However, we now begin to see that defining some limit to the system—although superficially 'convenient', indeed usually necessary to render laboratory or numerical experiments tractable—can overwhelmingly influence the way we view the system behaviour.

It is also critical that we permit interaction with the environment outside the system to occur without excessive constraint. This is quite in contrast to the traditional approach to experimentation whereby the environment is rigidly constrained, perhaps allowing only one of many potential variables to vary. It is this constrained, closed-system approach which has promoted and reinforced the Newtonian concept of matter as inert.

Quite in contrast, I argue here that geological processes are characteristically open-system processes in the presence of a flux of energy and matter. This flux, driving the feedback reactions which are typical of geological processes, leads geological systems away from equilibrium. We will see that rocks consequently evolve by a process called 'self-organization—they are active, not passive. This concept, following Ilya Prigogine's concept of active matter, is first and foremost a view of stirring natural beauty. Additionally, from the utilitarian viewpoint and in terms of society's exposure to geological hazards, the new concept holds the promise of major strides in engineering.

Hydraulic fracturing

Oil and gas wells are like upside-down drains. In poor quality oil and gas reservoirs, the hydrocarbon drains slowly to the well. This reduces the value of the reservoir because many wells are required to achieve economic rates of production from the field as a whole. In extreme cases, the hydrocarbon is left in the ground, the accumulation being deemed uneconomic.

Partly by accident, shortly after the Second World War, a process called hydraulic fracturing was discovered. This involves cracking the reservoir around the well and can be used to increase the flow rate. Whilst the process is expensive—up to perhaps one million pounds sterling including associated testing in the North Sea—the increase in flow rate is far from marginal. Typically, the flow-rate is more than doubled. The reason for this increase is that the effective surface area of the wellbore is increased by a factor of about 10 000. Before fracturing, flow at the interface between the low permeability reservoir and the wellbore experiences a choking effect similar to the entrance to a London underground station during the rush hour. After cracking, it is as if the width of the station entrance had been increased to, say, the length of Regent's Park.

In general, the crack forms perpendicular to the direction in which the natural geological forces in the reservoir are the lowest, which is therefore the easiest direction of crack growth. If we are lucky, and Nature is frequently kind to us in this respect, the forces, or stresses, are higher in the strata above and below the reservoir. In this case, if the crack attempts to extend upwards or downwards the tip of the crack is 'pinched' by these higher forces and so the fracture finds it easier to grow laterally. This is indeed fortunate, because upward growth would crack the seal which traps the oil or gas and downward growth would encourage production of the underlying water. Because the distribution of the stresses is so influential, these are an important input to the design of a hydraulic fracture.

So how do we actually crack the rock? Hydraulic fractures are engineered by pumping a viscous gel at high pressure into the well. The gel flows through small holes in the steel lining of the well and into the reservoir around the well. In the same way that a steam boiler will crack if the pressure gets too high, so will the rock around the cylindrical wellbore. Once the crack starts, the process becomes similar to the splitting of a log with an axe, with the axehead (Fig. 1) replaced by a wedge of viscous gel.

Fig. 1 Fracturing wood with an axe.

In fact, just like the crack in this photograph, it is even believed that there is a volume near the tip of the crack which the gel 'axelhead' cannot penetrate. It can be inferred that this volume fills with reservoir fluid (such as oil) which is then displaced back into the rock as the gel front advances during the next stage of crack growth.

The process is rather inefficient in that most of the gel is lost through the walls of the fracture into the reservoir. However, once the crack has grown sufficiently, suitable particles can be added to the gel to fill the fracture so that it cannot close after the pumping stops. Provided the operation has been successful, the particles form a very permeable pathway for the oil or gas which subsequently enters the fracture and travels to the wellbore. The gel is designed so that its viscosity greatly reduces after fracture closure, either by chemical reaction or in response to the high temperature of the reservoir. This allows the fracture to become free of gel when oil and gas production is resumed.

The net result is an elliptically shaped 'pancake' of permeable material approximately centred on the well. This provides both a large surface area open to flow from the reservoir and very little resistance to the flow once the hydrocarbon has entered the fracture. Typically, fracturing requires high rates of pumping and large volumes of fluid and propping materials. This necessitates the application of a great deal of horsepower and materials storage capacity (Fig. 2) to achieve the required geometry.

Fig. 2 Aerial view of a hydraulic fracturing operation.

An R&D programme conducted in the 1980s resulted in advances in hydraulic fracturing which had saved £275m by the end of 1992. These advances were recognized by the Royal Academy of Engineering in 1993 and resulted in the memorable experience of a presentation of the MacRobert Award for engineering innovation by HRH the Duke of Edinburgh jointly to Dr Paul Martins (a member of the Royal Institution) and myself.

With the single exception of the programme of experiments described below, the fracture designs assumed the reservoir rock to be a Newtonian entity—passive, inert, responding in a manner that is repeatable and precisely predictable given a knowledge of the imposed forces and Newton's laws of motion. Thus this programme represented but one example of the vast numbers of engineering operations which have successfully assumed the materials involved (usually non-geological) to behave in a Newtonian manner.

Behaviour of reservoir sandstone in the laboratory

The need arose to determine the stresses for our fracture designs. The conventional method of determining rock stress (force per unit area) involves monitoring the pressures required to drive a very small hydraulic fracture. This method, however, is not appropriate for most oil industry applications. First, the cost is too high in terms of rig time. Second, any suggestion to perforate the well below or above the reservoir is unlikely to be popular, especially if the purpose is then deliberately to crack the cement and rock behind the casing! Consequently, we sought a method which we could normally use without interfering with well operations. Rock core was chosen as the basis for our stress determination programme. This is a cylinder of rock obtained by using a hollow drill bit and is quite often available at no extra cost, having been collected for other purposes. The big challenge was, and in part still is, to relate the stresses in the rock from which the core is taken to measurements we can make of core behaviour. Thus arose the motivation for a programme of experiments on rock core.

The key: internal forces and their relationship to external forces

Imagine the cylindrical volume of rock in the reservoir about to be 'cored' by the advancing drill bit. Two categories of force act on this stick of rock. First, stresses act on what is shortly to become the outside surface of the rock core. These are termed external forces, and are no longer applied to the core after it has been separated from the reservoir by the advance of the hollow drill bit. The second category exists both in the reservoir and piece of rock which has been separated from the reservoir by the advance of the hollow drill bit. Forces belonging to this category are termed internal forces. If we could quantify the relationship between the internal forces in rock core, which we can observe in the laboratory, and the external forces which exist in the reservoir, we would be able to describe the force distribution which has so much influence on the geometry of a hydraulic fracture. Indeed, this would be a very substantial contribution to the subject of rock mechanics, which so often relies on little more than an educated guess of the value of this most critical parameter.

A typical specimen of sandstone core (about 10 cm in diameter and 10 cm in height) comprises about a million grains of sand cemented together to form a cohesive piece of rock. The internal forces in the core, which are prevented from being released by cementing minerals, the interlocking of the grains, and friction, arise through a wide variety of processes. One of the simplest is a cementing together of the stressed sand grains in the subsurface by a precipitating fluid. Figure 3 illustrates this process. This represents a pot in which we have placed a mixture of wet plaster of Paris and some squash balls (only a few of which are shown in this figure). If we imagine that the balls are squashed by a piston until the plaster has hardened, the plaster will prevent the balls assuming their original circular shape when we remove the piston (the action of which simulates the external forces in our reservoir). However, as the

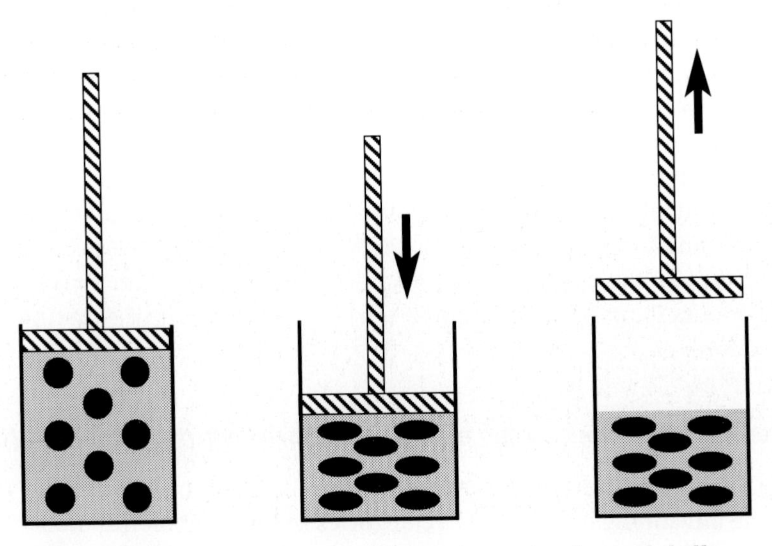

Fig. 3 Development of internal forces in a mix of squash balls and plaster.

balls attempt to relax as the piston is withdrawn, opposing forces develop in the plaster. Thus the plaster is stretched whilst the balls are squashed. Correspondingly, these internal forces develop in real rock long before it is cored by the drill because the conditions in our reservoir incessantly change through geological time, for example by change of burial depth.

No forces are present on the outside of our plaster-and-balls composite (except of course atmospheric pressure), but a complex balance exists inside. Quite how complex this is was first brought home to me 20 years ago on Mortlake station. I had been able to imagine the complexity which can arise from all the irregularities of sand grain and cement shape and properties, and a long and complex geological history. At Mortlake, however, I was shown the result of a numerical simulation of an immensely simplified representation of a real sandstone—simply a collection of perfectly round elastic disks of identical properties, devoid of any cement and subject to uniform external forces. I used to work with Peter Cundall who pioneered this modelling, and we were travelling together that day when he produced his latest results, a subsequent example of which is shown in Fig. 4. Note the complexity of the internal forces, despite the uniformity of the particles and their properties and of the applied stress. The internal forces are far from uniform and are transmitted in 'chains' through the assemblage of disks, some areas carrying high stress, some low. The thickness of the line of force chain is proportional to the magnitude of the force. This complex network of force chains prompted us to refer informally to this type of computer output as 'brain diagrams'.

I hope that you will now be able to appreciate that the apparent simplicity of a piece of sandstone is deceptive. A complex network of forces is present inside such a sandstone core sample, which is a product of its history.

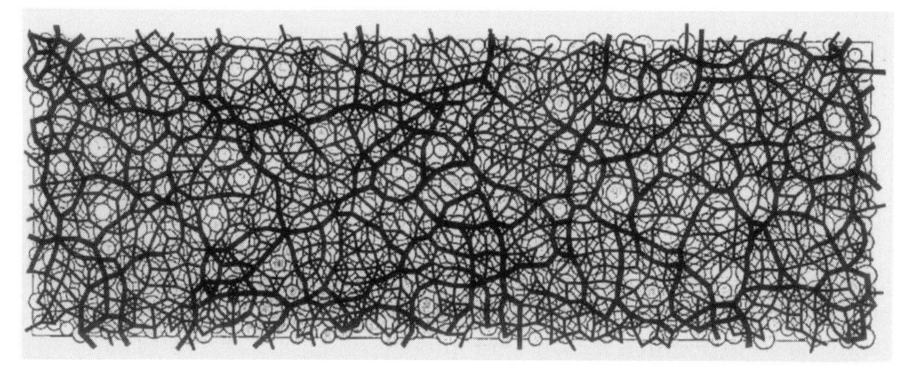

Fig. 4 Lines or 'chains' of force in Peter Cundall's model of stressed elastic disks.

Testing the assumptions of hydraulic fracture designs in the laboratory

Before we investigate the sandstone further, we will note some of the assumptions made as a basis for the successful hydraulic fracturing programme discussed earlier. These assumptions were incorporated in our computer model used to design the fractures and analyse the results of previous fracture treatments. These assumptions, which I shall go on to examine one by one, included the following:

1. The rock always behaves in a brittle manner, at least at the macroscopic scale.
2. The rock is elastic.
3. The rock is thermoelastic.
4. The rock is passive (Newtonian).

What do we mean by macroscopically brittle? This I can easily demonstrate by taking a hammer to a sample of the sandstone core and breaking a chip from the end of the core. We conclude that the sandstone is brittle when we observe the fragmentation that occurs on impact. But how does the core behave afterwards in response to part of the rock being detached and therefore becoming a new shape and size? To answer this question we need to carry out a more controlled experiment in the laboratory. Here we can carefully detach a disk of rock from the end of the sample using a diamond saw and monitor any subsequent changes in the size or shape of the rock using strain gauges. (This sawing process is also somewhat analogous to the propagation of a hydraulic fracture across the core sample.) We use electric resistance wire gauges that are bonded to the rock. When the rock changes shape the wire changes shape so that the electrical resistivity of the wire changes. The resistivity changes can be measured with great accuracy and calibrated for conversion to dimensional changes. Thus any changes in the dimensions of the rock can be accurately and automatically monitored.

Figure 5 shows the result of one such experiment. Let us examine the dimensional changes to determine whether the rock behaved in a brittle manner. Immediately we can see that the rock slowly expanded after the sawcut. It appears to have flowed! Looking more closely, we can see that the gradual expansion of the sandstone, eventually becoming larger by about 1/50 000 of the size it was before the sawcut, occurred over a period of about one day. All we did was saw the end off the core—there was no external force applied to the core afterwards to cause it to expand. So, whilst at first the sandstone appeared to be brittle, fragmenting under the hammer, a closer look revealed a quite different behaviour.

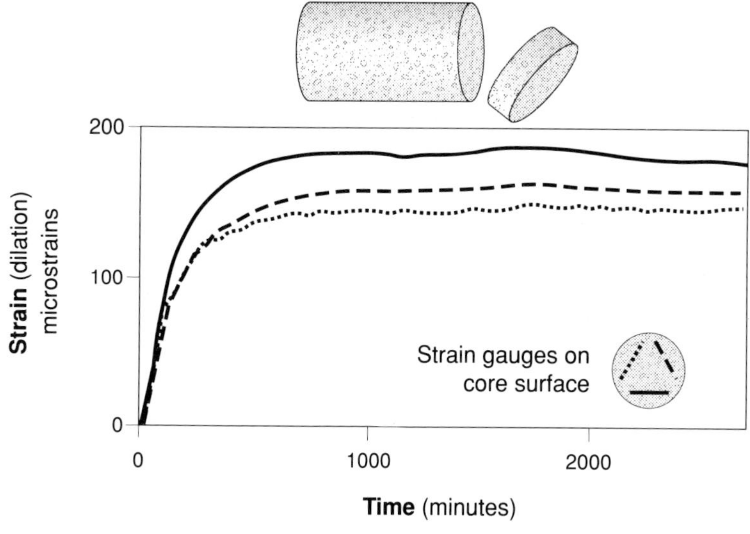

Fig. 5 Flow of a sample of sandstone after removing a disk of rock from the end of a core by means of a diamond saw.

I will emphasize here that the behaviour of materials which we observe, from which we develop our concepts of their behaviour, is crucially dependent on how we choose to view them, that is, dependent on the experimental test conditions and types of observation we choose to make. First the sandstone appeared to be brittle, then, when we looked more closely, we observed flow. So what should we expect if we choose a different type of observation? Indeed, to emphasize that what we see is a function of how we look, we can place small transducers on the rock surface which are able to measure acceleration. The outcome is that as the rock gradually expands after the sawcut, we observe microearthquakes, typically occurring at a rate of a few per minute. These microearthquakes represent microcracking. Now we are seeing a brittle behaviour again, but very localized: the length of the cracks might be about the same size as the diameter of a sand grain. Thus, observations of one type reveal a flow of the overall core as it slowly and permanently grows larger. Deep inside, however, tiny cracking events reveal that the flow is in part achieved by brittle fracture.

Now I shall briefly examine the assumption that the rock is elastic. By this, I mean that up to a certain yield point the rock deforms reversibly: after we remove the applied load the rock returns to the same shape and size just like a squash ball does after it has been hit around the squash court. Having seen the results of the diamond sawcut you will probably already be sceptical of this assumption. However, we can further test our reservoir sandstone using a principle of mechanics called after St Venant.

This tells us that if we locally disturb a stressed elastic solid, such as by drilling a hole in it, the effects will only be observed to a distance from the hole of a few hole dimensions (either the diameter or the depth of the drillhole, whichever is largest). Figure 6 shows a result from the only piece of core we have tested in this manner. This reveals dimensional changes at the other end of the core from that in which we drilled the small hole. The distance separating the strain gauges and the small hole was some 10 times the largest hole dimension (the depth of the hole in this particular test). We see that the rock sample did not respect St Venant's principle of stressed elastic solids; your scepticism is justified.

Are these reservoir rocks thermoelastic? This again refers to reversibility of shape and size changes but this time under the influence of temperature change. I must confess to never having checked it, but I feel safe in assuming that a squash ball gets larger when it is hit around the court because it gets warmer, because it is not only elastic but also thermoelastic. Numerous experiments have been conducted in which a clay-free reservoir sandstone sample is heated to perhaps 10°C for a few days and

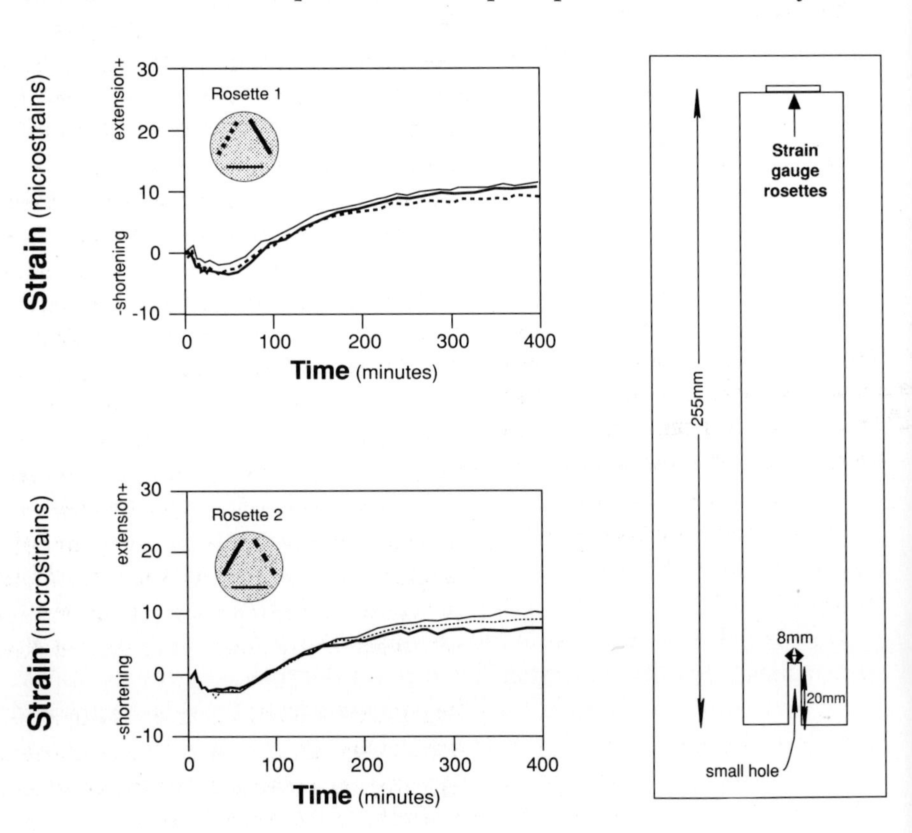

Fig. 6 A test for elastic behaviour.

Plate 1 Energy dispersal.

then restored to room temperature. The result is very similar to slicing the disk from the end of the core: the sandstone gradually expands and stabilizes in the expanded state. Again our assumption is not supported by laboratory experiments of this type—the sandstone is not thermoelastic.

The fourth assumption which we listed was that the rock is passive. It was viewed as Newtonian–passive and clod-like. We have subsequently tested this assumption by applying a load to the rock and observing the subsequent changes of size. Samples were coated with a silicone sealant and placed in a pressure vessel. Again a cycle was imposed, but this time of pressure instead of temperature. I think you would expect the rock to become smaller when we apply a positive force to the outside of the rock, i.e. attempt to compress the sandstone. Figure 7 reveals the result of one of these experiments. The rock does indeed contract when the pressure in the

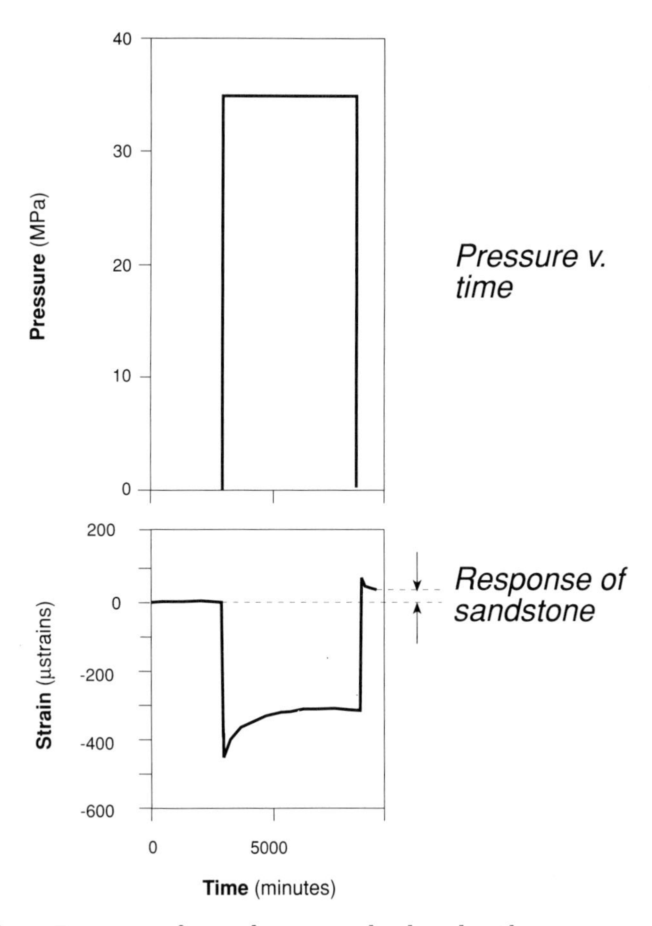

Fig. 7 Response of a sandstone to a load cycle. Above: pressure vs time; below: response of sandstone.

surrounding fluid is elevated. However, it gradually expands after this initial compression. This implies that the rock is exerting a force against the fluid, across the jacket of sealant, which exceeds the fluid pressure (almost 35 MPa or about 340 atmospheres). Indeed, after the fluid pressure is returned to the original value the rock appears to have permanently enlarged, the process of expansion occurring after some form of perturbation, evidently a rather general characteristic of this rock's behaviour. So, the rock was squashed for a few days but it ended up bigger than when we started. This is certainly not the response of a passive material. We can begin to suspect that the Newtonian concept of clod-like matter may not apply to this material. Our rocks may be active rather than passive.

What we see depends upon the way we choose to look

You will recall that we embarked on this experimental programme to investigate the possibility of describing underground stress from observations of core behaviour. We hoped to find a relation between internal and external forces. These experiments began to suggest that an interplay between the two does indeed occur and we found that underground stress directions correspond to the directions of core elongation we observe in the laboratory. We have begun to identify a different way of perceiving the rock behaviour which may be helpful in terms of the original objectives of our programme. But, perhaps you will have asked yourself, why did we make those four assumptions in the first place? The answer is that they comply adequately with conventional rock mechanics thinking as supported by conventional laboratory rock testing practice. Before giving you more details, allow me to fantasize for a moment in order to illustrate a central point of the topic under discussion here.

Immediately prior to delivering the Discourse upon which this article was based, I was confined to the lecturer's room in accordance with tradition. Had I wished to 'escape', rather than take the opportunity to collect my thoughts on Faraday's couch, I would not have been allowed simply to open the door and walk away. I would have had to resort to some unusual means of exit which stood some chance of remaining undetected. Recognizing that the window was too far above the ground to permit a safe exit, the possibility of escaping through the keyhole in the door might momentarily have crossed my mind in the desperation of the passing minutes. I would, of course, in the same instant reject this notion. Nevertheless, it would theoretically be possible for my body to be somehow forced through the keyhole. The experience would be fatal, but it could be achieved. Indeed, by using suitable pressure-measuring devices either side of the keyhole and some form of flow-measuring apparatus, a mathematical description of my flow characteristics could be identified. The resulting equation, given careful and sound scientific procedures of the standard we can expect in this Institution, would be a correct expression of the observations made—a description of my rheology (flow behaviour) under these conditions. It would, however, most emphatically not characterize my normal behaviour. I do not go through doors via the keyhole! Like the rest of us, I open the door, walk through, and close the door behind me. Strictly speaking, the mathematical expression derived by squeezing me through the keyhole would be correct—but it would be entirely unrepresentative of my characteristic behaviour.

I have subjected you to this fantasy to emphasize that it is absolutely critical that experimental conditions allow us to perceive that behaviour of the thing or process we are investigating which truly represents those conditions which we are trying to understand. We can get quite different results—a description of walking through a door or of flowing through a keyhole in my fantasy—both of which are strictly correct but only one of which remotely represents the characteristic behaviour. Our assumptions of rock behaviour did not come from fiction—such behaviour is easy to show in the laboratory if we tailor our experiments appropriately. (Incidentally, it would also be unreasonable not to point out that the description we used seemed largely adequate for purposes of fracture design).

There is an essential difference between the experiments recorded here and those more conventionally performed in the subject of rock mechanics. Conventionally, the rock is exposed to a prescribed externally applied condition throughout a laboratory (or numerical) experiment, such as a constant rate of loading or constant rate of displacement of the platens of the test machine. By contrast, the experiments reported here consisted of exposing the rock to some form of disturbance, such a sawcut or heating cycle, and then observing the outcome. The sandstone was not forced to comply with some externally imposed load or displacement throughout the test; after an initial disturbance it was free to *interact* with the environment beyond the boundary of the sample. In essence, the behaviour of the rock was observed more as an open system, whereas conventional rock testing typically represents closed system conditions. Our samples were free to experience the transfer of energy to (e.g. by heat transfer or acoustic waves) or from their environment and of matter (the samples expanded beyond their original boundaries). An obvious and highly relevant question is: which is most representative of natural geological processes?

The behaviour of open systems

Figure 8 illustrates an open system. As a result of the work of pioneers such as the Nobel prize-winner Ilya Prigogine, and the availability of digital computers, interest in open systems has greatly expanded during the last few decades. Some of the characteristics of open systems now recognized are as follows: energy dispersal, irreversible processes, non-equilibrium, and macroscopic order arising in response to a random flux of energy or matter through the system.

The observation that energy tends to disperse is embodied in the Second Law of Thermodynamics. This law expresses our everyday experience that we can expect coherent (ordered) motion to be completely convertible to incoherent (disordered) motion, but not the reverse. In closed systems, the spontaneous, or natural, creation of order from disorder is extremely improbable. We much more commonly expect disorder to develop spontaneously from order; a drop of ink in a beaker of water naturally disperses throughout the water but ink dispersed in water does not collect together to form a drop; the small dog in our cartoon (Plate 1) might wish he could organize all the energy he has just dissipated by frictional heating and convert this to work (coherent motion) to give him an extra boost to escape the large dog, but no dog has been known to do this. Whilst we would certainly not expect to encounter a situation in which the second law is violated, a crucial condition of our observation that energy disperses, i.e. incoherence forms spontaneously from coherence, is that we refer to closed systems. I will return to this point shortly.

The Second Law reminds us that we should expect irreversibility to be

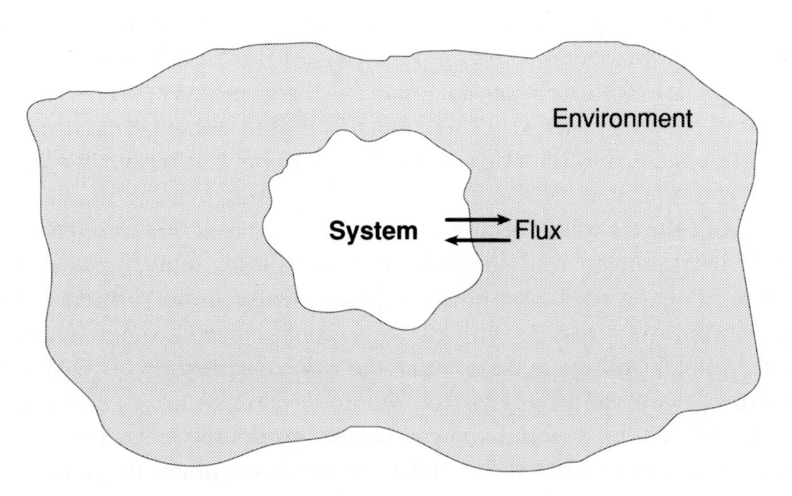

Fig. 8 Illustration of an open system.

commonplace. We can also use our reservoir sandstone to illustrate this. The 'brain diagram' (Fig. 4) revealed the complexity of a force distribution in a two-dimensional system of elastic disks. We can expect the force distribution in the sandstone—which is three-dimensional, of varied properties, and composed of irregular shapes—to be at least as complex. We can readily infer from our observation of shape and size changes, and of microearthquake activity, that the forces within the approximately one million sand grains evolve during the 24 hours or so after a sawcut experiment such as that shown in Fig. 5. In part, the force changes to which each grain is subjected are converted to heat, such as by frictional sliding between grains during the microearthquake activity we heard of earlier. Imagine that heat is emitted by grain A and naturally dissipated. The neighbouring grains try to expand and thus in turn apply forces back to grain A—and so on for a vast number of interactions throughout the rock. Minute cracking forms new free surfaces in an instant, in response to force changes, at the same time as a host of changes are occurring both around the crack and in other parts of the rock. Every change experienced by an individual grain affects the neighbouring grains.

Self-organization

Open systems can be driven away from thermodynamic equilibrium by a flux of energy or matter through the system. There is an increasing recognition that in this non-equilibrium state it is possible for macroscopic order to arise spontaneously, as I shall illustrate by means of examples. This order is manifest as structures termed dissipative structures because they develop at the expense of the dissipation of energy. The process of order formation is graphically called 'self-organization'. But, you will say, what about the Second Law? I will reply to you that it must be argued that this does not contravene the Second Law. You may think that this is 'passing the buck', but we must infer that the order which is created in our system is accompanied by the creation of a greater amount of disorder outside the system.

Systems which respond non-linearly to a disturbance, such as sandstone, are most readily driven far from equilibrium. The behaviour of such 'dynamical systems' depends not only on the present state (e.g. present force distribution) but on the previous state. Such systems display the much-popularized characteristics of fractal geometry and chaos. In this non-equilibrium state they also characteristically fluctuate. The microearthquake activity which follows after our sandstone has been subject to a sawcut confirms that the activity varies from place to place in the sample apparently with no regular pattern either in time or in space throughout the rock core. This is a form of fluctuation of our non-equilibrium system, but in reality only a crude representation of the myriad of changes too small to locate with our insensitive measuring system. The core, originally circular, becomes increasingly elliptical each time we disturb it. The ellipticity is a form of macroscopic order and is broadly repeatable as indicated by the degree of consistency of the direction of maximum expansion (Fig. 9).

We have now come a long way. We started with the perception of

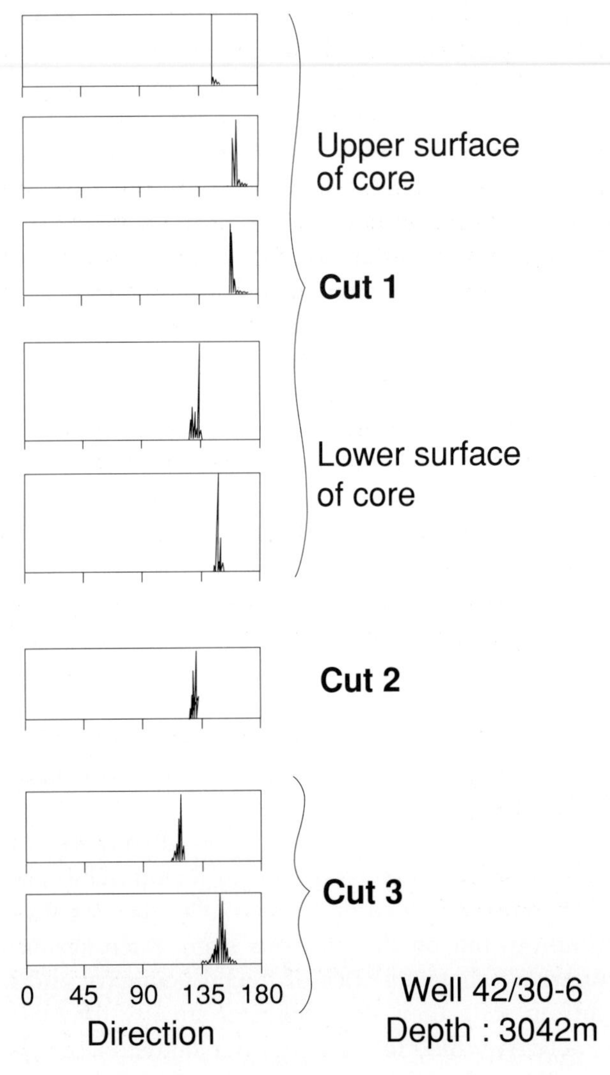

Fig. 9 A form of macroscopic order development: the elliptical shape of the sandstone is orientated in a repeatable manner.

sandstone as passive and clod-like. Then we saw that we can easily create experimental conditions in which the rock most certainly does not respond in a manner which is predictable simply by applying Newton's laws of motion to the precise values of the external forces. In essence, in common with many other systems which are also capable of self-organization, the rock is able to create *order* when subject only to a *random* flux of energy (such as, in our case, gentle heating). We find that we can view the rock as active.

The easiest way to illustrate this process of self-organization is by example. Figure 10 shows a fluid between two plates, heated from below. This reproduces a famous experiment but we can equally think of a pan of water on the stove at home. At first the molecules of water are randomly distributed within the confines of the pan. When the water gets very hot, order develops in the form of convection cells, illustrated in Fig. 10. This is the response to a random flux of energy, in the form of heat, through the water. The convection cells appear to be a response of the system which is 'designed' to facilitate the flux of energy through the system. It appears that the system is organizing itself to increase the rate of dissipation of energy—the convective rolls comprise a 'dissipative structure'.

We often see structure in clouds. Figure 11 is an example taken from an aeroplane. This is another example of self-organization. Returning to earth, but still in the realm of fluid mechanics, observe the turbulence next time you are near a river in flood. The precise location and the times of formation and decay of each of the whorls and eddies keep changing (fluctuating) unpredictably. When the river is low the flow is much more uniform; when in flood this structuring, or order, aids the dissipation of energy. Finally, we need only to turn to our neighbour. Yes, we too are dissipative structures, exhibiting self-organization. Indeed, if my ambitious

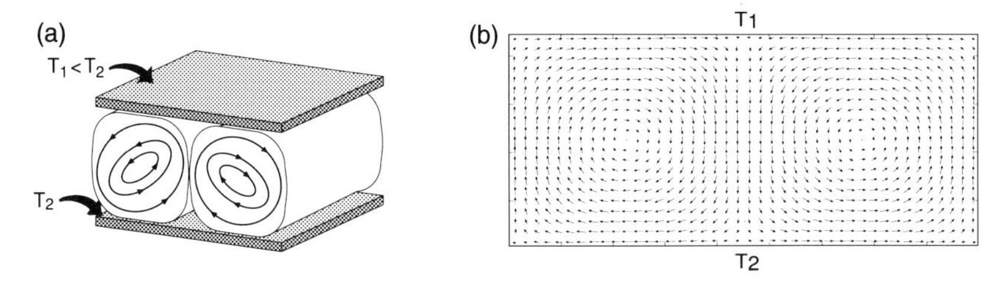

Fig. 10 Convection cells are an example of macroscopic order developed in response to a random flux of energy through an open system. (a) Convective rolls in a fluid heated from below. (b) The microscopic molecular order of the convective rolls. This example shows the fluid between two plates, but organized convection structures also form when water is heated in a saucepan.

Fig. 11 Clouds are another example of a dissipative structure.

attempt to adequately express these wonders in this article is at least partly successful, the distinction between what you normally accept as living and dead will no longer be as clear as before you started reading.

We have emphasized the neighbour-to-neighbour interactions in self-organizing systems such as our sandstone. Contrast this with the past, when we would have been more likely to focus on the substance, such as details of the material forming the sand grains. For a final example of a self-organizing system, we can program a computer to model *only* the interactions and completely ignore the substance, allocating to each location only a state of being dead or alive (on or off). This is the game of 'Life' invented by the mathematician John Conway. A two-dimensional matrix of cells can be used as an array of possible locations at which 'birth' or 'death' can occur. Each location has eight neighbours. The three rules of interaction, simulating the influence of population density on reproduction and survival, are simply as follows: birth occurs when a cell has three neighbours; death occurs either by isolation (when a cell has two neighbours or fewer), or by overcrowding (with more than three neighbours). These are very simple rules. What, if anything, do you expect to see when we instruct the computer to cycle repeatedly through the array, checking neighbour–neighbour relations for births and deaths at each location?

The essence of this game is that the system evolves. Starting with a seed configuration, you would observe an evolving mixture of apparent randomness (as you perhaps expected), patterning (order), stability, and fluctuation. Yes, order and fluctuation can arise in this simple model of neighbour–neighbour interaction. Is that what you expected?

Geological processes and engineering applications of the new concept

Now let us turn to the utility of our new conceptualization of rock. Is it helpful? I began this article with man-made fractures used to help us extract a natural resource. I shall finish by considering natural fractures which are hazardous to us so that we have an engineering need to minimize the hazard. This involves an understanding of structures at the scale of the Earth's crust, much larger than our laboratory cores. Some of you may question whether we can extrapolate this far and that is a fair question which deserves an answer before we go further.

My colleague Dr Hagan and I have been continuously experimenting with reservoir sandstones for almost nine years. The tests have been designed to reveal the overall behaviour of the sandstones. We have not tested any more than a very few other rock types and so I must emphasize that we have no experimental basis to claim that all rock cores would behave in the way I have described here. Indeed, we would not expect to be able to record the changes so clearly in many other rock samples. Within the Earth's crust, however, it is easy to show that the conditions necessary for self-organization are widespread and that we can observe the same characteristics of non-equilibrium, non-linear systems as those of the evolving sandstone described earlier.

First, the crustal processes must occur under open-system conditions. We can check this condition by sketching a cross-section of the Earth and drawing on it a rectangular (or any other shape) box near the surface to define our system. All we have to do next is sketch the outward heat flow from the centre of the Earth to demonstrate that the system is inevitably open. Next, we must ask whether we should expect feedback relationships, in which case the system is also non-linear. To do so we turn our attention briefly to the rheology, or flow behaviour, of rock. Here we will find the feedback we were checking for, and appreciate that the system is non-linear.

Imagine we are for the first time entering the discipline of rheology, which is concerned with the deformation and flow of materials. If we are fortunate, we quickly encounter Reiner's second axiom of rheology, which is both concise and most instructive. This tells us that all materials possess all properties, including elasticity, brittleness, and plasticity,

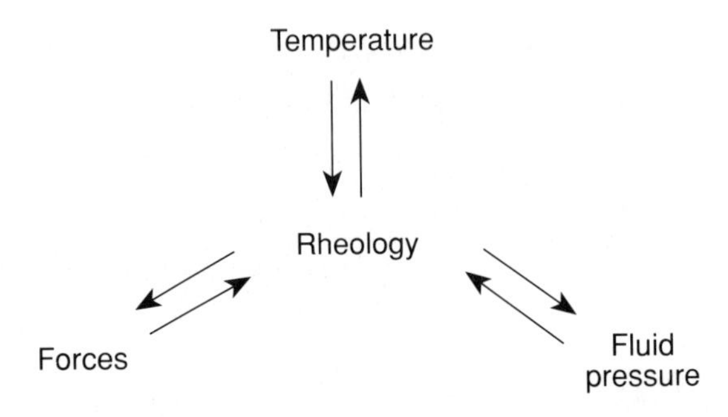

Fig. 12 The interplay of forces, fluid pressure, heat and rock rheology in the Earth's crust drawn to emphasize the fundamental role of the feedback relationships.

depending on the so-called environmental conditions. For example, if you lean against a window pane in your house, it will bend but return to the same shape when you move away. However, it behaves in a brittle manner when your son kicks his football through it yet when it was made it was heated so that it would flow into a sheet. Rocks will accord with Reiner's axiom in closed-system conditions. In the Earth's crust, by contrast, the interplay sketched in Fig. 12 prevails. Consider one example of this interplay. Not only does temperature affect the rheology of the rocks but the rheology also affects the temperature. Imagine that the rock cracks. The cracks will facilitate heat flow through the rock and the temperature of the rock will consequently change. The rocks do not simply respond to a change of the 'environmental parameters', they change their environment. This realization first hit me like a thunderbolt, having previously encountered so much closed-system thinking, slavishly followed, it seems, both in the laboratory and in numerical models. Again we must conclude that the rocks are active, not passive.

I hope you are now satisfied that we can indeed extrapolate to the Earth's crust and we can now agree that we should expect to encounter self-organizing, non-equilibrium systems. In accord with this we might expect fluctuation, and in fact this is graphically demonstrated in any record of vertical crustal movements. Fluctuation is evident at all timescales. The geological record reveals many examples of subsidence followed by uplift during periods of tens of millions or hundreds of millions of years. The surveyor who returns to a traverse to repeat a geodetic survey (within of course the span of no more than a few tens of years) always encounters fluctuation and may find himself seeking in vain for a benchmark which has remained stable.

Natural cracking of the Earth can be an industrial hazard, environment-

Fig. 13 A consequence of the earthquake of 17 January 1994, Los Angeles.

ally detrimental, and a danger to human life. Rupture at the ground surface can break pipelines, crack building foundations, or damage roads. The magnitude of the hazard relates to the consequences, so we must be particularly careful in the siting and design of facilities such as nuclear power plants, nuclear waste disposal facilities, and high pressure chemical or hydrocarbon processing plants. The shaking accompanying earthquake rupture has created much loss of life and damage. Recently Los Angeles experienced more damage (Fig. 13), yet we still struggle to understand adequately the earthquake process. Traditionally seismology has regarded rock as a passive, Newtonian entity. Now there is an emerging interest in treating the crust as a non-linear system. Earthquakes may not be as predictable as we would ideally like because they are a manifestation of a non-linear dynamical system. Prediction of the time of an earthquake therefore becomes akin to forecasting the weather. Much more development is required before we can demonstrate the utility of the new concept in engineering terms. However, the potential of treating the Earth's crust as an open, self-organizing system seems immensely promising. I hope you begin to suspect that this also applies equally to disciplines other than seismology.

For a final example I return to where I started—hydraulic fracturing of oil and gas wells. The fractures were engineered by bringing to bear a large quantity of horsepower to enable us to force the rock apart and make it accept the propping agent. Our subsequent experiments then revealed that the rock is, after all, active and willing to respond to a small disturbance non-linearly by using its own internal energy. I would like to leave you with a final question. What do you think we might achieve if in the future we work *with* the rock rather than against it?

Acknowledgements

Joe Hagan conducted the core experiments reported here. I welcome this opportunity to express my thanks for his ceaseless good humour and easy nature. Informal conversations with Jim Berry and Jerzy Szymanski have been invaluable. Jim Berry kindly programmed the game of 'Life' as a basis for a video used in the Discourse and helpfully commented on the manuscript. Our understanding of rock behaviour would not have progressed so far without the expertise of John Shaw of Liverpool University, who invented a reliable method of determining the pre-drilling orientation of the sandstone core samples. I am grateful for the opportunity provided by BP, including a short sabbatical at ETH in Zurich. In struggling with the interpretation of the core behaviour, which has given so much pleasure, I drew upon the writing of many others. Finally, the forbearance of my family in tolerating my frequent disappearance, only to reappear muttering about the Second Law, dissipative structures, and self-organization, merits at the very least a mention in these acknowledgements.

TIM HARPER

Born 1943, took degrees in geology and engineering rock mechanics at the University of Southampton and Imperial College. After lecturing in the Engineering Department at Aberdeen University, he joined a firm of consultants, where he was part of a team required to predict the mechanical evolution of the geological environment of a nuclear power plant in the USA. In 1980 he proposed to British Petroleum an improvement of the technique of well stimulation known as hydraulic fracturing. The consequent research programme received the MacRobert Award for engineering innovation in 1992. In 1991 he set up a programme to characterize the mechanical behaviour of approximately 2000 cu. km of rock hosting a large oil accumulation in Colombia.

Materials in the fast lane

D. T. CLARK

Introduction

Formula 1 (F1) motor racing, as well as being an exciting and interesting sport, provides an interesting example of man–machine–environment interfaces and extremes of materials requirements. It is a field where the distinctive contributions of chemistry, physics, materials science, and engineering and their pivotal roles in enabling the science and art of the designer to be translated into a world-beating package are perhaps understated.

The rapidity of regulation changes and the strong influence of the weather and of other cars on the track, provide a challenging and testing regime for new materials, whilst the extremes of conditions experienced by the drivers (high *g* loads, high temperatures, rapid reaction times, etc.), provide the epitome of man–machine–environment interaction, in a high technology sport, which provides a vehicle not only for rapid technology development, but also for technology diffusion to a wider sphere.

With wide audiences (the 16 Grand Prix events held around the world during a season typically attract 4 million spectators, with TV coverage adding an additional 160 million) and widespread coverage in newspapers and specialist magazines, the F1 world excites strong emotions and two contrasting viewpoints might be expressed as follows:

Viewpoint 1

- a mindless, expensive game for overgrown school people
- SCALEXTRIX with sound writ large
- an environmentally disastrous use of resources
- a minority sport of no practical utility

Viewpoint 2

- a cost-effective, rapid vehicle for technology development/diffusion relevant to a variety of industries/markets
- a sophisticated entertainment requiring great mental and physical strengths of the drivers

- an unsophisticated entertainment with midget 'gladiators' who can drive

- an area of engineering and materials design where the UK leads the world
- the most cost-effective marketing/PR medium devised to date
- universal appeal

In the spirit of Faraday, in this short article, I evaluate the scientific evidence to show that only the second of the two views is sustainable.

F1 being the pinnacle of motorsport provides an exceptional testing ground both for materials and for materials systems. It has provided a rugged and rapid test ground for the development of 'battle-hardened' technologies many of which we see now in everyday road-going vehicles; examples of these technologies include **multivalve engines, turbocharging, radial tyre development, engine management systems, advanced engineering composites, fuel and oil development, traction control, semi-automatic gearbox development, and rolling road wind tunnel development,** all with considerable spin-off in other markets.

The Formula

The mission in F1 can be stated quite succinctly as **'To transport an individual in competition with others around a road circuit in a car defined by F1 FIA (Federation Internationale de l'Automobile) regulations as rapidly as possible over the defined race distance.'**

The overall shape of the F1 car is defined by design engineers, using the basic laws of physics (especially aerodynamics), and, as will become apparent, the complexity of the overall 'shape', is enabled by sophisticated chemistry to new materials, with very high specific and absolute properties.

The races themselves are held between March and November at a combination of custom-built tracks and street circuits spanning the continents. The majority of races are still held in Europe, the historic home of 'Formula' racing; with the noticeable exception of the USA, where the somewhat less sophisticated Formula of 'Indy' racing dominates, spectator and TV coverage is only really rivalled by the Olympic games, which are held only every 4 years. This high level of interest is partially explained by the fact that since so many people drive, or at least periodically ride in cars, there is an immediate affinity in the battle between man–machine and environment, to which people can relate.

It should perhaps be noted that the design of a typical racing circuit is very different from that of a normal road or motorway in terms of materials. Whereas a public highway has to cope with high traffic loadings and significant wear, etc.,. from high axle weight freight vehicles, the average race track only sees relatively modest use (and in any case a F1 vehicle is typically only half the weight of a family saloon), such that the thickness of the construction of the latter is typically half that of the former (38 cm vs 78 cm). Since grip is so important, however, the F1 track typically has 4 cm of bituminous concrete in the surface finish compared with only 3 cm in a typical motorway. Materials therefore play a very important role in facilitating optimum performance from the F1 vehicles in their interaction with the track.

It is salutary to note that motor racing in all its facets is now celebrating its centenary;[(1)] however, the origins of 'Formula' racing with its closely defined rules relating to driver, car, etc., date back to 1900 or so with the introduction in France of the so-called 'Gordon Bennett' races sponsored by the US newspaper magnate to stimulate international motorsport competition as illustrated in Table 1.

Formula racing has gone through several eras, where different nations were dominant, from the early days when France was the dominant force, through the era between the wars when Germany was pre-eminent, to the post-war era when first Italy and more latterly the UK has been defining the cutting edge of the technology. Whilst in the early days the direct relationship between road and racing cars was apparent (Plate 2a), the modern era has been one of aerodynamically designed single seaters where front and rear wings, and undertray design, contribute to the ability to generate very high downforce and hence cornering capabilities.

Plates 2b and c show two cars from the modern era both from the excellent Williams Grand Prix Museum which is one of the most complete in tracing design history in the F1 field; FW06 (Plate 2b) dating from 1978 is of alloy honeycomb construction whilst the latest FW16 (Plate 2c),

Table 1. The early days of Formula racing: the 1900–1905 Gordon Bennett races

Race	Features
First motor racing formula	Vehicle weight 400–1000 kg without driver, luggage, food, etc.
1900 Paris–Lyon (570 km)	Average speed 62.1 km/h Pan hard 24 HP
1905 Auvergne circuit (548 km)	Average speed 77.9 km/h Braiser 96 HP

the 1994 car, is a state-of-the-art carbon fibre/Kevlar/Nomex/aluminium honeycomb construction which will be analysed in a later section.

A further feature evident from Plate 2 is the modern-day phenomenon of advertising and financial sponsorship. The Williams of 1978 as well as displaying financial sponsorship from oil and tyre companies also shows an early example of sponsorship from a non-motoring organization (Saudia Airlines); the growth of media coverage, particularly TV, has seen an explosive growth of such sponsorship, which provides significant drive in financing the technology development which now drives the sport. (A competitive team now requires something like £30 m a year to mount a serious challenge at the Driver and Constructors titles.) The current Williams FW16 reveals the importance of sponsorship not only from automotive, oil/fuel/tyre companies but also the contribution made by tobacco companies such as Rothmans.

Whilst at ICI I was involved in the excellent collaboration between the company and Williams Grand Prix Engineering. In addition to providing a cost-effective way of publicizing the ICI roundel, F1 has provided a significant technology testbed for ICI products, as is evidenced from Plate 3.

The basis of the change in chassis/monocoque construction, exemplified by Plate 2, is encapsulated in Plate 4, where the move from the heavy, steel-based chassis of the early racing cars has been successively superseded first by lighter weight alloy/honeycomb structures through to the modern generation of stiff light carbon-fibre-based monocoque constructions with the engine/gearbox being bolted or adhesively bonded as stressed components of the total structure.

Figure 1 (the examples again being drawn from the Williams Grand Prix Museum) compares the alloy and carbon composite structures, the latter being exclusively used because of their overwhelming advantages since the early 1980s to the present day. The development of stiff, light damage tolerant driver 'cells' has contributed immeasurably to driver safety and even some of the massive accidents of recent years, (for example, Gerhard Berger in his Ferrari at Imola and Mauricio Gugelmin in his Leyton House March at Le Castellet during the 1989 season and Allesandro Zanardi in his Castrol Lotus at Spa Francorchamp in 1993), have seen the drivers rapidly back in action. The alloy/honeycomb technologies employed a decade ago would have almost certainly seen fatalities from such high speed 'shunts', bearing testimony not only to the greatly increased performance of the current-day vehicles but the greater driver protection afforded by advanced materials.

The safety factor requires individual certification of the monocoques to be used by any team in a racing chassis and the FIA monitors both crash

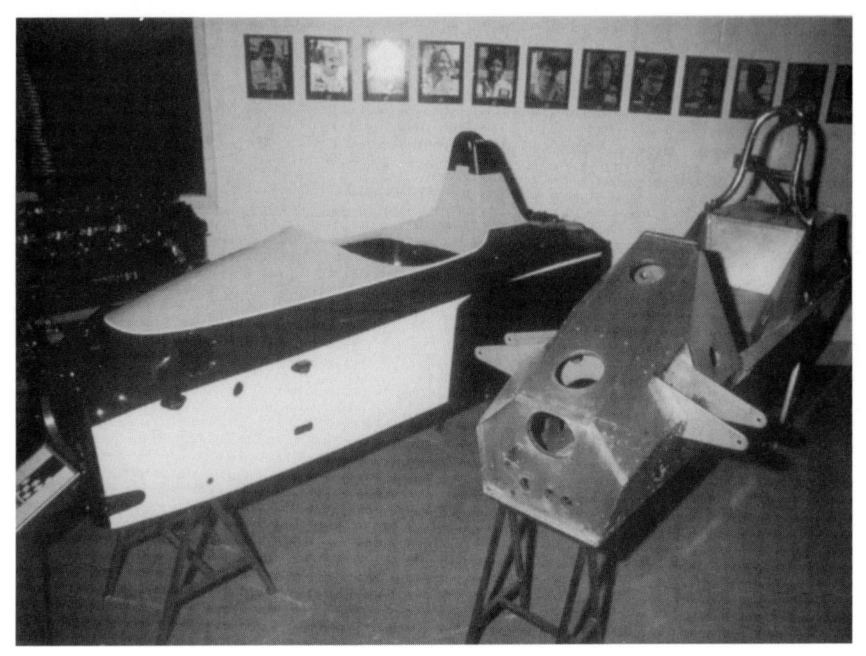

Fig. 1 Comparison of alloy/honeycomb and carbon-fibre F1 'tubs'.

testing and loading tests so that each F1 racing car is scrutineered at the track before and after a race, particularly in regard to the FIA approved certification panel which is 'sealed' to the structure.

The development of F1 chassis particularly in regards to improvements in stiffness and lightness has placed significant new demands on the drivers, a point we shall consider in some detail later, and the increased *g* loadings now require levels of fitness very much higher than 20 years ago; it also means that, relatively speaking, F1 is a young persons' sport. This contrasts with, say, Indycar racing where F1 drivers well past their 'sell by date' can still remain competitive. (Indycar vehicles are typically 50 per cent heavier than their F1 equivalents and the tractability of the higher revving engines for F1 give much higher acceleration rates; also, the use of carbon/carbon braking systems in F1 cars compared with the steel disks still employed in Indycars, typically reduces braking distances to less than half that in Indycars so that the *g* forces exerted during acceleration and braking are typically much higher in F1).

The impact of technology on cornering forces is illustrated in Table 2.

The F1 rules and regulations are detailed year by year in the FIA 'yellow book' and are generally aimed at slowing cars down and or restricting the technology advantage enjoyed by the leading teams. For example, the

Table 2. Development on cornering forces

Decade	Development	Cornering force (g)
1950s		~0.6
1960s	Wider tyres, rear engine	~1.0
1970s	Winged cars	~2
1980s	Advanced composites (stiffer, lighter structures) advanced tyre developments	
Present	Higher specific engine outputs	~4

1994 season sees the introduction of regulations aimed at restricting microprocessor-based driver aids so that 'active' suspensions, traction control, and advanced ABS braking systems have been banned, and compulsory refuelling stops have been instituted, all aimed towards making the races less of a procession behind the most technologically advanced teams.

There is therefore, a continual tension between regulation and technology with the latter generally winning. This is perhaps best illustrated by comparison of two of the leading cars of their year, namely the last Williams Honda of 1987 and the 1993 world champion car powered by Renault. The Honda V6 turbo-charged vehicle produced 950 bhp in race trim and with the banning of turbochargers by the regulators (since 1988) the normally aspirated Renault V10 in race trim typically produced close to 800 bhp. The improvements in total design (aerodynamics, semi-automatic gearboxes, traction control, active suspension, etc.) are such that it is estimated that in a back-to-back race the 1987 car would be some five laps behind the 1993 car.

Whilst the scientific basis for F1 total design is now very significant, there are still sufficient areas of 'black art' for the intuitive designer still to be important and, despite the rapid advances in computer-aided design and routine use of advanced 'rolling road' wind-tunnels (most often tested on roughly half-scale models), it is still possible for knowledgeable organizations to come unstuck in spectacular fashion. Recent examples of this genre include the Porsche excursion into F1 engine design[2] and the Lola attempt at F1 chassis design,[3] the latter contrasting dramatically with Lola's success in the somewhat less sophisticated Indycar field (Nigel Mansell's Newman–Hass Indycar championship car of 1993 was designed by Lola[4]).

The human dimension in F1 should not be underestimated, as will be evidenced by a later section on the driver; the regulations, however, have placed a premium on total teamwork; examples of this include times of

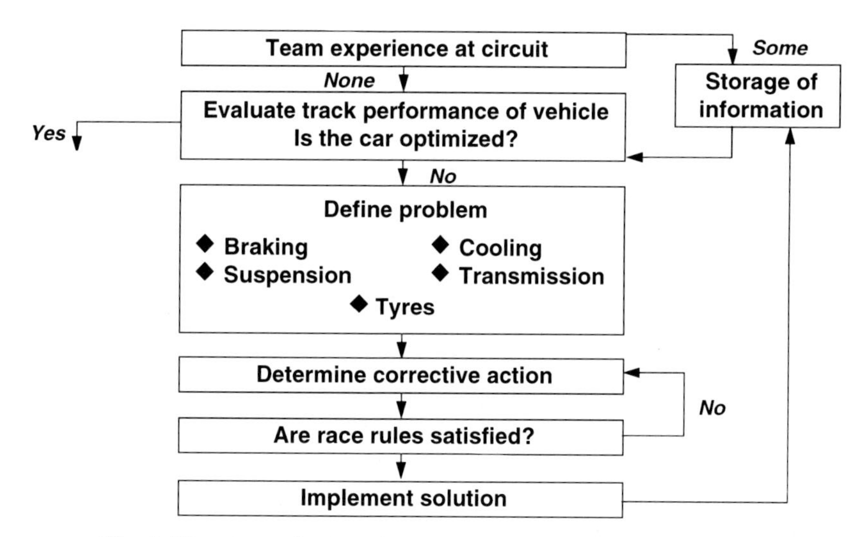

Fig. 2 F1 set-up; the learning process.

5 sec to complete a four-wheel change, 1 h for an emergency engine swap, half an hour for a gearbox change, 10 sec for a nose cone change, and 10 sec to dispense 100 litres of fuel in 1994 refuelling stops.

A further influence of the regulators is the attempt to stop the learning process at specific tracks, reinforcing the competitive advantage of the most technologically advanced teams, with their ability to make full use of on-board real-time monitoring and two-way telemetry with the pits to optimize the chassis set-up of the engine, etc., at each track as a function of the environmental conditions. (For 1994 only one-way telemetry and two-way voice communication are allowed.)

The typical 'quality' process of learning by measurement employed by the top F1 teams is illustrated in Fig, 2.

Whereas in the 1980s teams with large budgets could afford extensive testing at all of the F1 tracks, in advance of the 3-day race regime (2 days' practice and timed practice sessions to determine grid order, followed by one race day), each team must now nominate an official F1 track for their practices; practices are then banned at the other tracks, prior to the races, for the teams for which that track is not their nominated test track. This still provides competitive advantage for those teams who have logged all of the test data from previous years, since changes to track layout are relatively infrequent (for example, the introduction of chicanes at Silverstone again aimed at slowing the cars down!). Again it is this continual tension between scientific and technological development and regulation which makes the sport so interesting and exciting.

Road cars and F1: a comparison

As we have noted, in the historic development of Formula racing the relationship between road and race cars has continued to diverge, although some exotic sports cars can still trace their origins to the F1 race field rather directly. Examples might be the **Ferrari Testarossa** of the 1980s, the first to be designed with the F1 layout of rear engine and rear mounted oil and water radiators, with the striking comparison of the F1 airducting pods being simulated by the elegant side strakes of the Ferrari. The space-frame ladder structure, however, still relates to F1 of the previous generation, whilst the new **McLaren F1** road car genuinely uses some of the weight-saving technology associated with current generation F1 cars. Even so, with the highest power to weight ratio of any current road-going car, the McLaren still looks puny by comparison with the real thing: the power to weight ratio for the top ranking F1 cars (**Williams, Ferrari, McLaren, Benetton**) would still be nearly three times higher.

It is instructive, therefore, particularly in relation to the demands on the driver, tyres, etc., to look at a comparison of typical family saloon, an exotic sports car, and an F1 car, this is done in Table 3. A particularly striking comparison is the distance travelled in the time it takes an F1 car to travel 1000 m from rest. This is shown in Table 4. The striking acceleration and deceleration capabilities of the F1 vehicle, for example in braking the car from 170 to 70 m.p.h. (in taking a tight chicane after a straight), in 2 sec, place severe requirements on driver fitness. This is considered briefly below.

The driver

Mankind has been racing wheeled vehicles almost since the discovery of the wheel; some of the characteristics of an F1 driver include:

- exceptional reaction times
- effective communicator between machine, designer, and engineer
- peak fitness
- asymmetrical muscling (since most circuits are clockwise)
- preferably small and light
- ability to operate at a variety of levels
- fearlessness

The typical reaction times of an F1 driver are one-third those of the average person in the street; top drivers (Nigel Mansell, Ayrton Senna, Alain Prost, Michael Schumacher, Jean Alesi) are capable of reading the

Table 3. Comparison of Formula One and other cars

	Formula One car	Exotic sports car	Family saloon
Typical tyre temperature (°C)	100	40	25
Typical downforce at maximum speed (%)	300	90	70
Top speed (mph)	~200	~185	~115
Typical rel. weight (at rest)	0.5	2	1
Power/weight ratio	14	2	1
Wheels	Mg alloy	Al alloy	Steel
Chassis construction	C fibre to bonded to drive train (engine/gearbox)	Steel ladder space frame Steel/Al/ composite cladding	Mild steel Monocoque
Typical Cd[1]	~1	0.3	0.4
Fuel tank type capacity (gallons)	Rubber/Kevlarcell 40	Thermoplastic 25	Thermoplastic 10
Range (miles)	~200	500	300
Fuel consumption	Av. speed 130 mph 4 mpg	70 mph 24 mpg	50 mph 35 mpg
Engine life before rebuild (miles)	500	150 000	70 000
Bore/stroke	1.6:1	1.05:1	1.1:1
Intake and exhaust	Unfiltered; no baffle boxes	Filtered; two baffle boxes	Filtered; two baffle boxes
Catalytic convertor	×	√	√
Gear box	6/7-speed semi-automatic	5-speed manual	5-speed manual/ automatic
Oil system	Radiator dry sumped	Radiator dry sumped	Wet sump

[1]Cd = aerodynamic resistance coefficient.

Table 4. Acceleration of different cars

Car	Distance travelled in 12 sec from standing start (m)
Family saloon	120
Ferrari Testarossa	330
Williams-Renault	1000

road, (i.e. have good awareness of what is on the road and at the side of it) at 170 m.p.h. with a level of feedback equivalent to the average person travelling at 50–60 m.p.h. Since the drivers routinely pull 4 *g* laterally during practice and racing, and since most circuits are clockwise, the body as an 'active' materials system tends to develop asymmetrically, particularly in the neck muscle region, and the occasional anti-clockwise circuit (e.g. Imola) often leaves a driver with a sore neck after the race.

Since F1 vehicles effectively have zero compliance in their suspensions (most readily appreciated in watching F1 by studying the cars as they come into the pits for a tyre change, where jacking the car up leads to no relative movement of the wheels relative to the car (see comparison with road cars in Table 8), the driver's ride is fairly rugged, and communication between body and chassis, important from the point of view of controlling the man–machine interface, often leaves the drivers with both bruised elbows and hands after the race. The physical demands are reinforced by the relatively hostile environment of the driver compartment with no creature comforts (no upholstery, and little protection from the elements), and the natural first reaction of the drivers at the end of a race is to drink water to replenish the typical 7 lb of weight lost during the race.

The need for drivers to be light should be self-evident, since there is no current equivalent of the weight handicap used for jockeys in horse-racing (although this has been used to produce competitive racing in the German Touring car series) and 10 kg of additional driver weight can translate into a lap time penalty of 0.2 sec. Since space is at a premium inside the narrow cockpit, tall drivers are at a disadvantage: Gerhard Berger, the Austrian Ferrari pilot, is over 6 feet tall, whilst those who are above average height and happen to have large feet have a double disadvantage. In this respect the outstanding performances of Damon Hill at Williams during his first Grand Prix season in 1993 were all the more commendable.

The developments in car–pits telemetry arising from the silicon revolution, and indeed F1 as a hostile testing ground for battle-hardened electronics, have moved apace over the past few years, and in the pits at an F1 race the space taken by data-logging/analysis/telemetry is almost as large as that devoted to the cars themselves. Plate 5 shows, for example, part of the telemetry taken during Nigel Mansell's so-called ultimate lap at Silverstone in practice for the British Grand Prix in July 1992. Of particular interest are the lower two traces showing the lateral and vertical *g* forces during the race. Since the fastest lap inevitably means that the driver is on either the throttle or the brakes (or, for those skilled in left foot braking, on both at the same time), the *g* forces continually in acceleration or deceleration and both laterally and vertically exceed 4 *g*. To put this in perspective, 4 *g* represents the sort of *g* loading experienced by the supine space shuttle astronauts at take off, or the sort of *g* loading one would experience at one point during a circuit of a state of the art roller coaster. By comparison, the F1 driver has to control his vehicle in relation to others, he needs often to plot laps in advance of his overtaking moves (often needing to take advantage of the hole in the air slipstreaming effect so evident when watching the races), and the combination therefore of physical and mental effort is unparalleled in any other sphere of activity, adding the strong human interest which makes F1 such an exciting spectator sport.

The supreme fitness of the drivers is evidenced by their ability to compete in a range of sports; for example, a year or so ago Alain Prost competed as an invited guest rider in one of the mountain stages of the prestigious cycling event, the Tour de France, finishing an impressive 37th in his own right in the stage.

Lest one be accused of being sexist it appears to be a biological fact that the combination of physical and mental strength and stamina required to be a competitive F1 driver, perhaps akin to a combination of a top class tennis player and a 5000 m middle distance runner, seem to be best suited to the male of the species, and in recent years there have been no competitive female F1 racing car drivers. By contrast the number of female members of the teams involved in engineering/design, in construction/assembly, in the computer-based data logging and analysis, and in the PR/marketing and financial aspects of F1 have been increasing, as have the number of researchers in materials science, etc., of relevance to the field.

Materials requirement for F1

Formula 1 with its extremes, in terms of materials demands, is an excellent practical engineering test-bed for materials science. Since performance is everything (sponsors are difficult to come by if cars are not consistently on the winner's podium), emphasis needs to be on 'cost of engineered solution', rather than cost of materials mindset that characterizes a lot of engineering thinking. In some areas the requirement will be for ultimate properties, while in others the drive is towards a specific property, *viz.* property per unit mass of material. In designing structures that are optimum for performance, the question of anisotropic engineering design raises its head; that this issue can only effectively be addressed by 'composite structures' is one reason amongst many why carbon-fibre based systems are so important to the field.

The basic principles behind anisotropic structures with fibre and matrix with differing properties are learned at an early stage as are the design principles behind 'tough' damage tolerant structures where the interface is a sizeable fraction of the total structure. Creating a new interface dissipates a lot of energy. Wood fractured across the grain demonstrates this, and the complex topography and high surface area contribute to the high fracture toughness of some woods when loaded in the appropriate direction. A natural composite like wood has an anisotropic 'grainy' structure and its mechanical and fracture properties are different in different directions. We chop wood most readily by going along rather than across the grain; we also learn that laminating wood in layers with differing grain directions in successive layers make structures (plywood) which are damage tolerant, whilst anyone who has tried to chop chipboard (essentially a random composite with a wood-chip filler and a resin matrix), will know that it is a very tough material. On a nano and ultimately molecular scale the same design features obtain in the man-made composites that are such a feature of F1.

Many design considerations have to do with weight. For example,

- FIA specifies a minimum weight of 500 kg empty
- there is a power:weight ratio—an extra 10 kg weight adds 0.2 sec
- it is useful to be able to alter the distribution of weight, e.g. when turning corners per lap
- for reciprocating moving components, e.g. wheels, brake disks, pistons, inertia must be considered

Therefore the emphasis has been on materials with very high specific properties, such as metals (titanium, aluminium, magnesium alloys), composites (carbon fibre, Kevlar), and 'honeycomb'/foam structures like Nomex AL PU, with densities as low as 0.04.

If we consider the 'chassis', (an important component of which is the driver monocoque), we can consider this in simplified terms as a tube-type structure which needs to be stiff (since **minimum distortion under load gives spatial control of interaction with the road); of minimal weight; and strong and damage tolerant.** The requisite mechanical efficiency for a structure can be subdivided into a component relating to the intrinsic properties of the material, and an extrinsic component related to the ability to fabricate complex shapes and the properties of these shapes. For materials in general there is a strong interaction between molecular order/morphology and shaping and design and hence the properties of shapes. Indeed, one of the great scientific challenges for the future is to move chemistry/materials science closer to design/processing science.

Table 5. Properties of different materials

Material	Stiffness	Density	Toughness	Shape
Wood	√	√√√	—	√
Metal	√√	√	√√	√√√
Carbon fibre composite	√√√	√√√	√√√	√√√
Ceramic	√√	√	—	√

Therefore carbon composites are extensively used where temperature is not a dominant issue. Density can be reduced by foamed/honeycomb structures.

Table 5 gives a schematic overview of some desirable properties of materials and the ability to produce shapes. It should be clear from this as to why carbon-fibre-based systems in general hold such a dominant position. Figure 3 compares specific properties of steel and aluminium with those of carbon-fibre composites. In many senses man has taken a leaf from nature's book in designing stiff, light, strong structures optimum for function, by combining materials in composite form.[5] In the case of carbon-fibre composites, for example, we take a stiff, strong, lightweight fibre (some 7 μm in diameter, roughly one-tenth of the diameter of a human hair), which is somewhat brittle, and combine it with a relatively tough resin matrix to produce a material which is stiff, strong, light, and

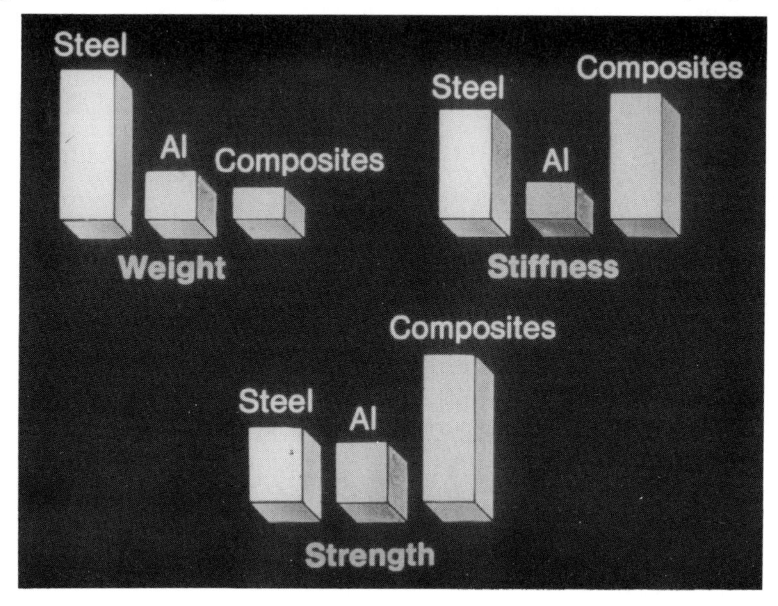

Fig. 3 Comparison of specific properties of carbon-fibre composite, steel, and aluminium.

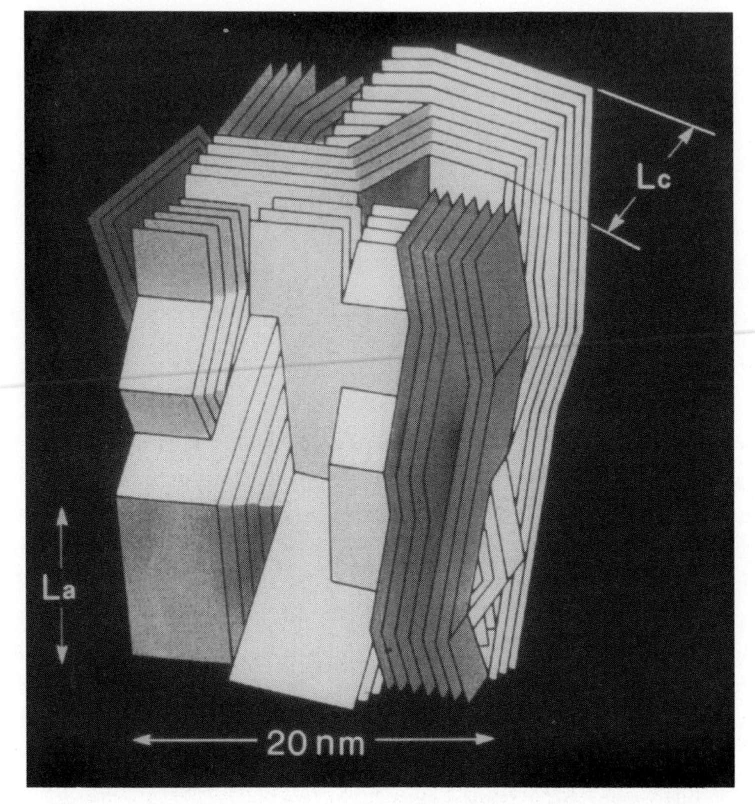

Fig. 4 Schematic representation of a section of a carbon-fibre.

tough and can be fabricated into shapes with differing properties in different directions. A schematic of a carbon fibre is shown in Figure 4.

If we take as an example (Aromatic Polymer Composite; ICI–Fiberite trade name for PEEK–carbon-fibre composite), APC-2, one of the highest performance carbon-fibre composites currently available, a typical 10 cm^3 sample contains 160 km of carbon fibre, providing an interface some 3.5 m^2 in area. A typical F1 car, therefore, if all the carbon fibre used in its construction were laid end to end, would reach from the Earth to the Moon! Figure 5 shows some of the properties of APC-2.

The typical components currently manufactured from high performance polymer composites (almost all at the present time thermoset in nature) are shown in the exploded diagram in Fig. 6. In considering fibre placement to give the required degree of specific stiffness in a given direction, attention is drawn during the 'lay-up' stage to the type of structure that is required. Table 6 lists some specific examples.

In a short article such as this I do not have the time or space to go into great chemical detail but current F1 cars are largely constructed from

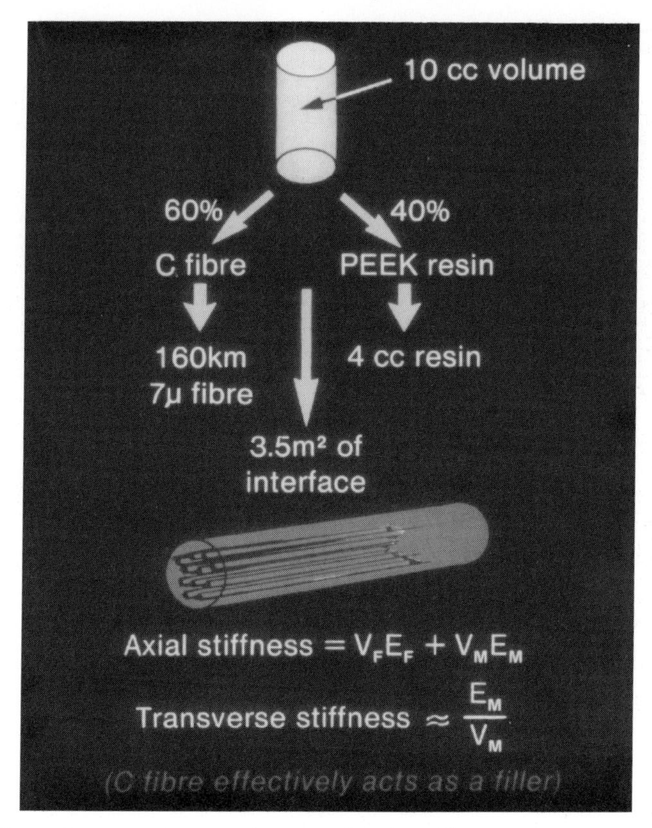

Fig. 5 Some properties of a typical carbon-fibre composite (APC-2)

resin systems which are slightly more sophisticated versions of domestic araldite® thermoset epoxies. These are relatively brittle and typically have a tenth of the fracture toughness of thermoplastic based composites such as APC-2. They do have the advantage, however, that the pre-preg sheets (uniaxial, cross-ply laminates and wovens) with the uncured resin still at the 'tacky' oligomer stage have tack and drape leading to easy hand lay-up in or around moulds using scissors, scalpels, and hair dryers. The typical sequence including the ovening process is shown in Plate 6 for the construction of an air intake/engine cover for a Williams F1 car.

By comparison, the higher performance thermoplastics in which all the chemistry is complete are stiff and require new high temperature moulding techniques.[6] Interestingly enough, therefore, F1 does not yet use the most advanced carbon fibre composites available, partly because there is an existing investment in process technology and partly because the competititive advantage at this stage is marginal. Whilst this is true for

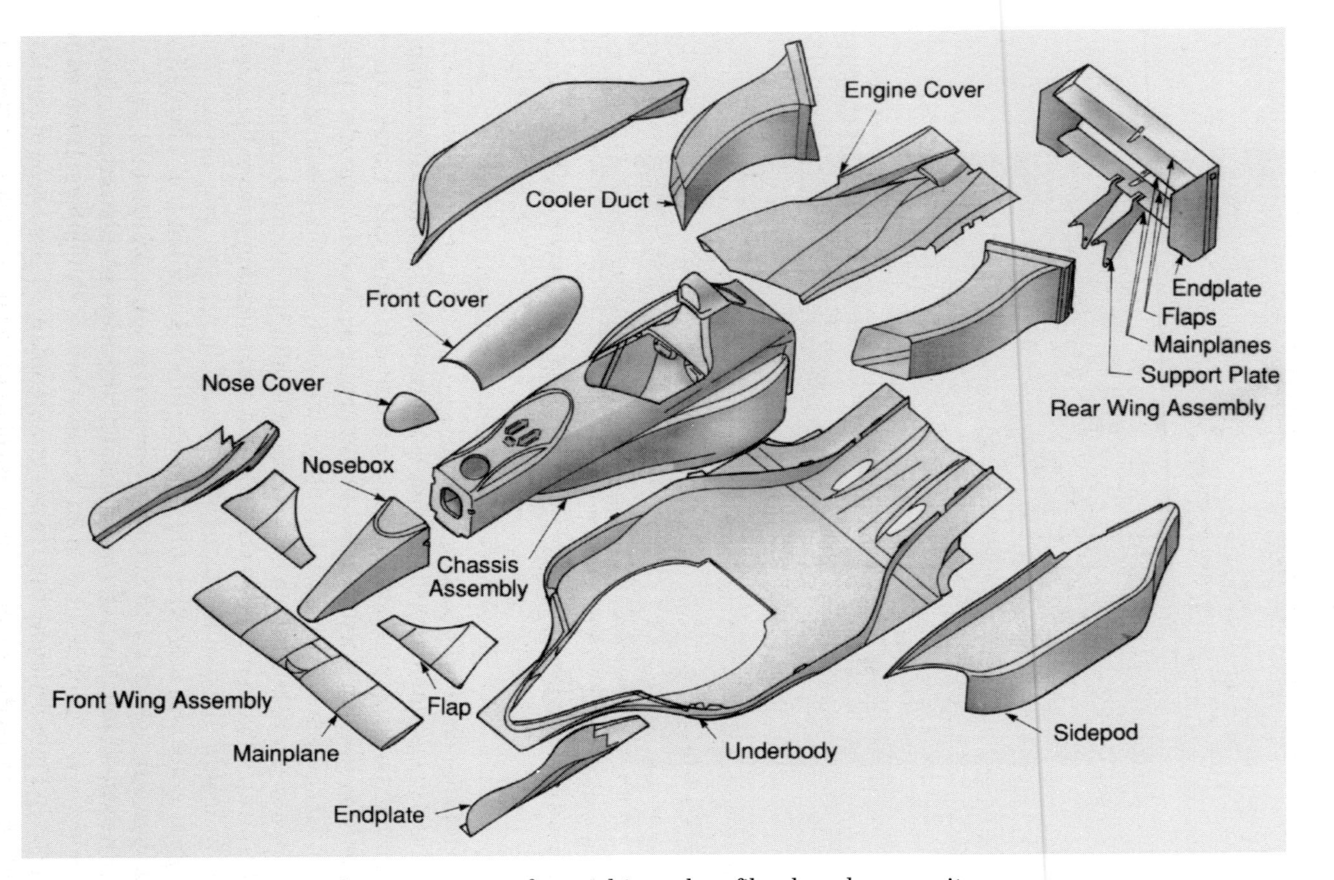

Fig. 6 Some components of an F1 car manufactured in carbon fibre based composites.

Table 6. Examples of specific carbon-fibre placement for different structures

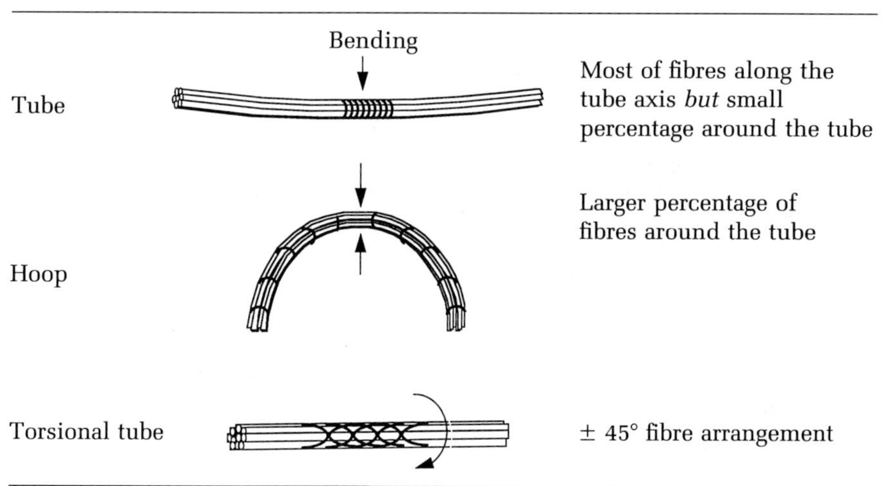

Structure	Fibre placement
Tube	Most of fibres along the tube axis *but* small percentage around the tube
Hoop	Larger percentage of fibres around the tube
Torsional tube	± 45° fibre arrangement

passive components of the structure, such as the 'tub', wings, side pods, engine cover, etc., once moving, reciprocating drive train components are considered, the fracture toughness required mandates a thermoplastic carbon-fibre component. In a collaboration between the author, his ICI team and a group from the composites centre at Imperial College, gear selector forks for the Williams semi-automatic gearbox have been evaluated based on APC-2. Figure 7 shows a comparison between the carbon fibre composite component and the forged steel component.

Although carbon-fibre composite costs 10 times more in materials than the steel component, the processing advantages mean that the carbon-fibre composite component can be produced more cheaply than that from steel. It is stiffer and lighter than the metal component and the tribological performance far exceeds that of the metal; this is important in street circuit races like Monaco where during the race the some 1500 gear changes occur, on average one every 3 sec, and where gear selector wear is an issue.

By control of molecular architecture it is now possible to produce a new generation of toughened thermoset materials, where the development of a co-continuous (molecular knitting, warp and weft molecular arrangement) morphology, with a large area of interface between a thermoset and typically 20 per cent thermoplastic phase gives materials with five times the fracture toughness of traditional aerospace epoxies but with the advantage that their tack and drape characteristics are virtually identical. Although the research underpinning the development of a

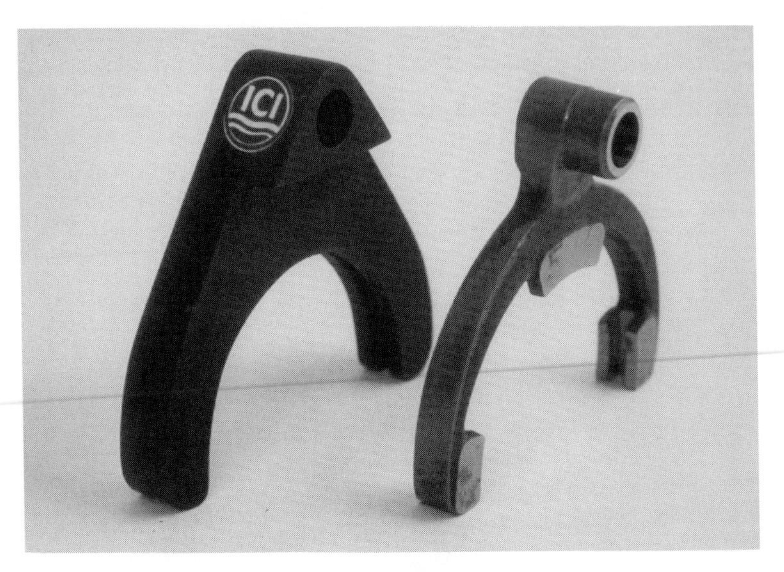

Fig. 7 Comparison of APC-2 and forged steel gear selectors.

generic route to molecular level toughening of composite matrices was only initiated in the author's ICI laboratories five years ago these materials are now entering extensive use not only in F1 but also in aerospace in general. This approach[7] is outlined in Fig. 8.

Fire-resistant clothing

With F1 vehicles carrying up to 40 gallons of fuel, and with the ever-present possibility of a crash and fire, F1 rules are very strict in terms of low flammability, self-extinguishing underwear and overwear for the drivers and track marshalls. The F1 chassis is made essentially of carbon, and the flammability of typical carbon-fibre composites is significantly lower than that for magnesium, for example, as used for the wheels and gearbox housing/engine sump. Materials systems with a very high carbon to hydrogen ratio are typically self-extinguishing, a fact which is familiar to anyone who has been frustrated to see their barbecue briquettes extinguish even though they have been primed with lighter fuel of much higher hydrogen to carbon ratio.

The natural material cotton is excellent for the underwear for the average car driver, but for the F1 driver low flammability Nomex is prescribed not only for underwear but also for socks, gloves, balaclava, and for three-layer box quilted oversuit worn by all drivers, and for the fire protection suits worn by the marshalls.

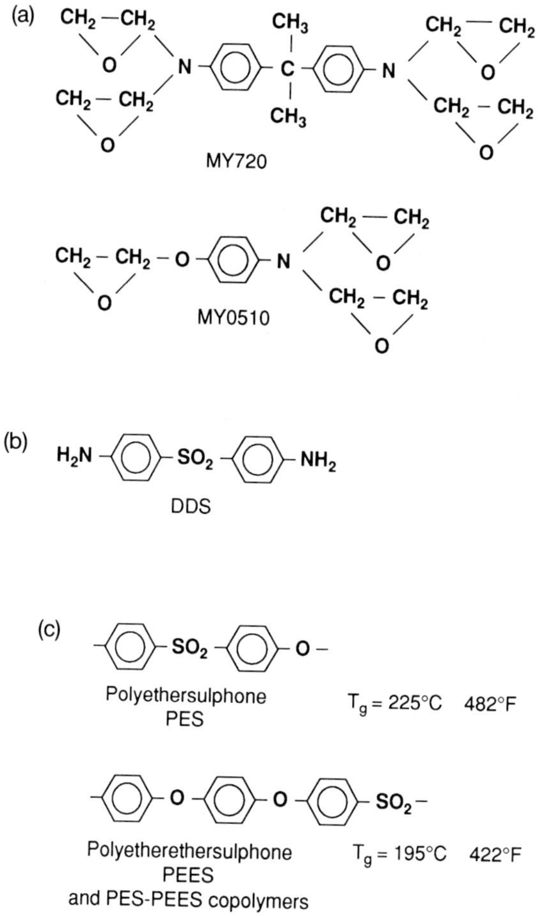

(a)

MY720

MY0510

(b)

DDS

(c)

Polyethersulphone
PES $T_g = 225°C$ 482°F

Polyetherethersulphone
PEES
and PES-PEES copolymers $T_g = 195°C$ 422°F

Fig. 8 Molecular level strategies for co-continuous morphology development in toughened thermosets. (a) Typical epoxy resins; (b) typical curing agent; (c) thermoplastics used.

As is evident from Table 7, Nomex and Kevlar, developed by Dupont are close relatives as aromatic polyamides. The meta-links of Nomex lead to a more disordered (more soluble) structure which is easier to spin fibres from whilst Kevlar acts at appropriate concentration as a lyotropic liquid crystalline material which is difficult to process into the fibre forms used to such good effect in F1 composite applications. Nomex is also easily processed into honeycomb forms used in the deformable nose cone of the F1 vehicles.

The investigation of the fire-resistant properties of fabrics has been particularly developed by Dupont with their ‘Thermoman thermocouple-

Table 7. Structures and properties of Nomex and Kevlar

Polymer	Repeat unit	Properties/uses
Nomex	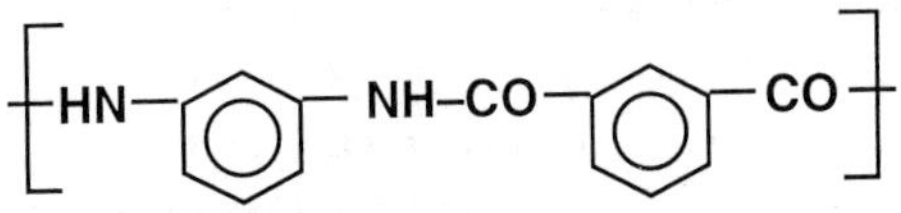 ⁅HN–C₆H₄(meta)–NH–CO–C₆H₄(meta)–CO⁆	Similar to nylon but heat resistant. Used for flame-resistant clothing, e.g. underwear, gloves, driver's suit, balaclava, etc. Honeycomb used for nose cone structure.
Kevlar	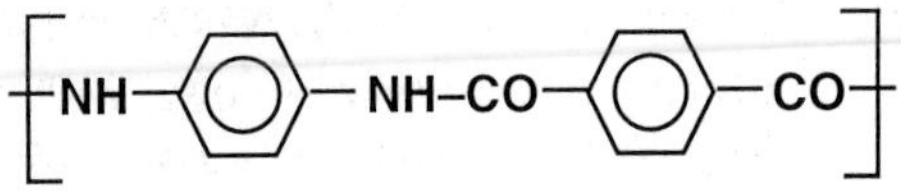 ⁅NH–C₆H₄(para)–NH–CO–C₆H₄(para)–CO⁆	Very strong fibres used in construction of composite body shell, helmet, and tyres.

wired mannequin'. The requirement to withstand burning fuel at a temperature of 1200 K for 12 sec is a demanding one and the application of Nomex-based textiles and their effectiveness is demonstrated by the low degree of fire injury risk experienced by F1 drivers. For example, in the horrendous crash suffered by Gerhard Berger in 1989 at Imola, the ruptured fuel tank left the driver sitting in neat fuel; however, the rapid response of the fire marshalls and his protective clothing left him, remarkably, unscathed from the crash.

In addition to its significant role in F1, Kevlar also sees increasingly widespread use in the general automative field in cylinder head seals, power transmission belts, clutch linings, heat-resistant hoses, brake linings, and tyres. It is also increasingly seen in consumer goods such as high performance skis, canoes, etc. as well as its use in aerospace in general.

The tyres

In terms of shaped materials systems tyres represent one of the most complex (and secretive areas of F1. The essential function of the tyres are to **interact with the road to produce the forces necessary for the support and movement of the vehicle (acceleration, braking, cornering, etc.), and to cushion the vehicle against road irregularities (much more important than for road-going cars with their compliant suspensions).**

Indeed, it has been stated that '**the areas of contact between tyres and road are the very front line trenches in the furious battle between space and time**'.

(a)

(b)

(c)

Plate 2 The progression of 'Formula' racing cars: (a) an early example; (b) FW06 (1978) Williams; (c) FW16 (1994) Williams.

Plate 3 ICI products on the Williams F1 cars in the recent past.

Key: (A) Autocolor: automotive paint system; (B) Metron: actuator devices for fire extinguisher and life support systems; (C) BCF halon: fire extinguisher compounds; (D) Propathene: polypropylene for brake fluid containers; (E) Weav/rite and Tool/rite: composite bodywork materials; (F): Fluon: PTFE in bearing surfaces; (G) ICI Racing Fluid: high performance brake fluid; (H) Kephos: metal treatment system; (I) Polyurethane chemicals: energy management foam for seats; (J) Carbon/PEEK composite: gear selector fork.

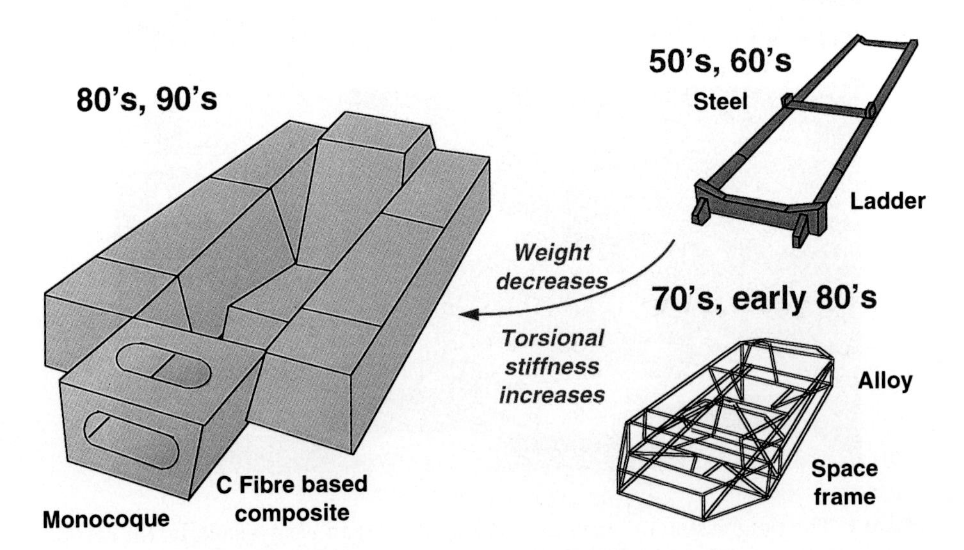

Plate 4 Progression in materials/design for F1 chassis.

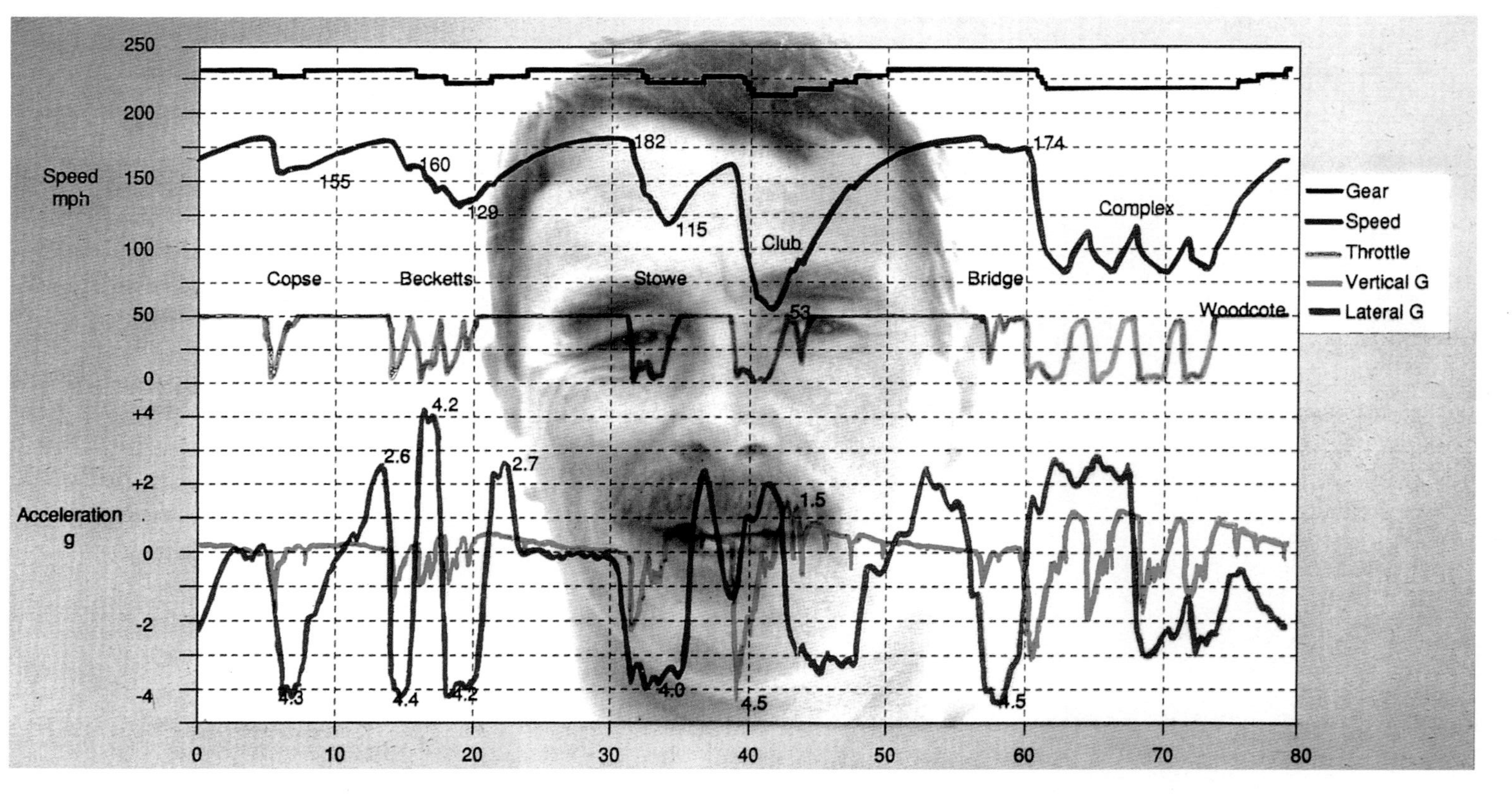

Plate 5 Telemetry from Nigel Mansell's ultimate lap at Silverstone on 10 July 1992.

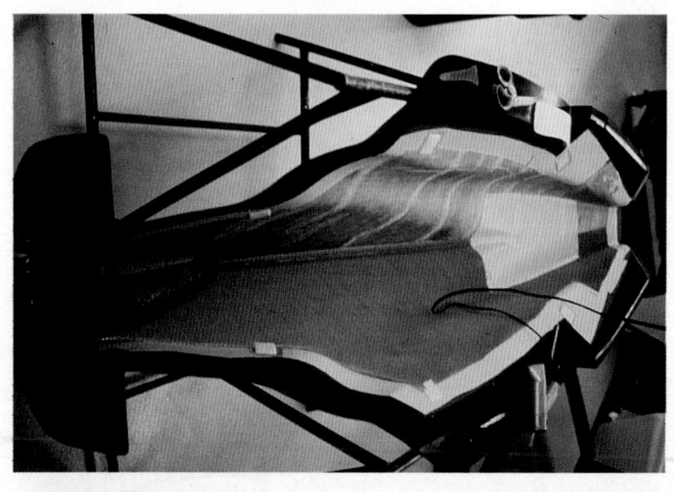

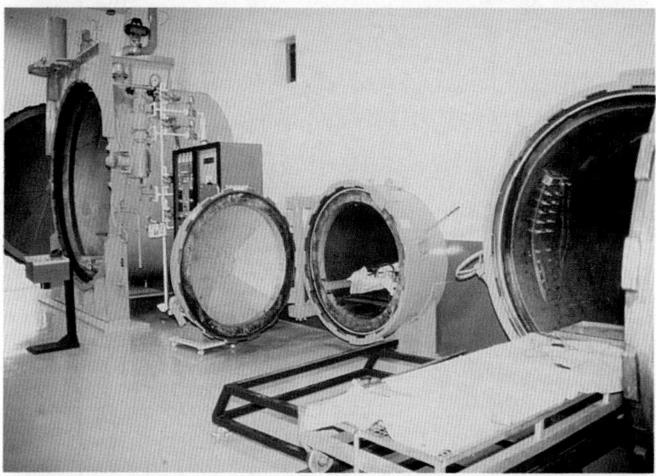

Plate 6 Hand lay-up, oven cure cycle, final engine cover/air ducting for a F1 car.

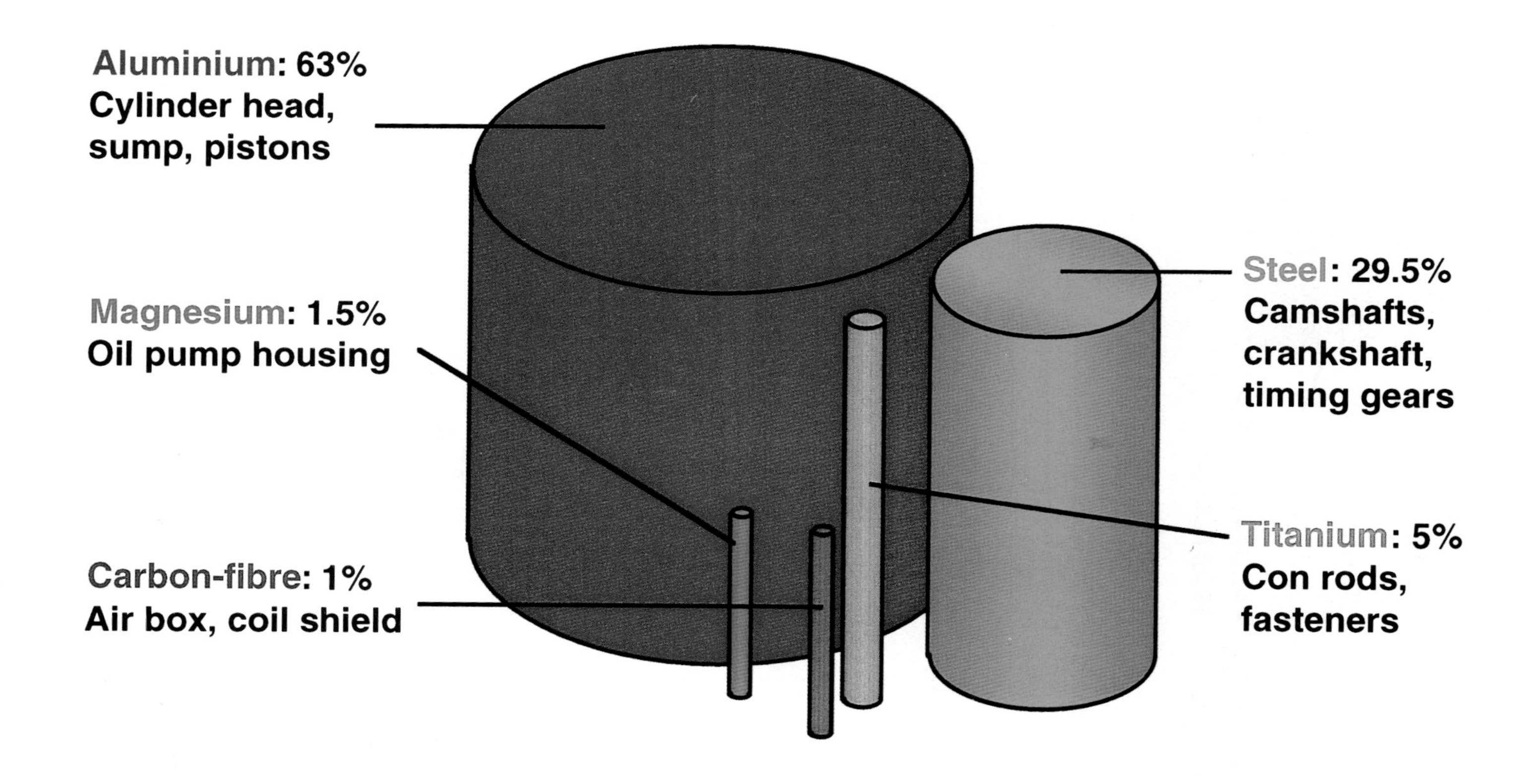

Plate 7 Typical materials make up of an F1 engine/drive train. (Except for certain parts that need to be made from special materials, F1 engines are principally made from aluminium and steel.)

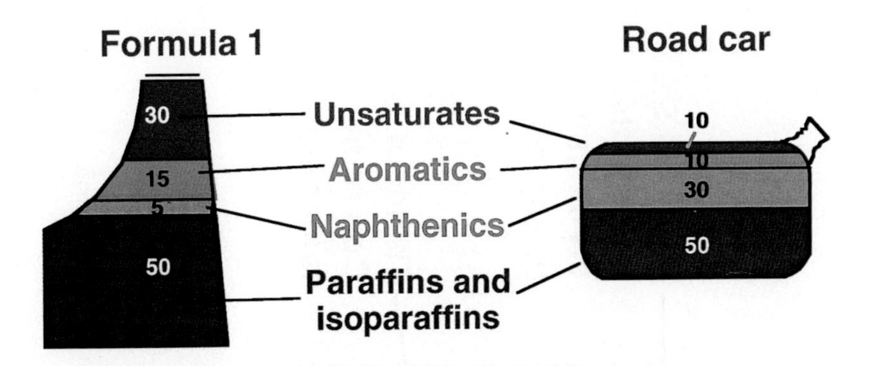

Plate 8 Comparison of chemical composition of F1 (1991) and road-going pump fuel.

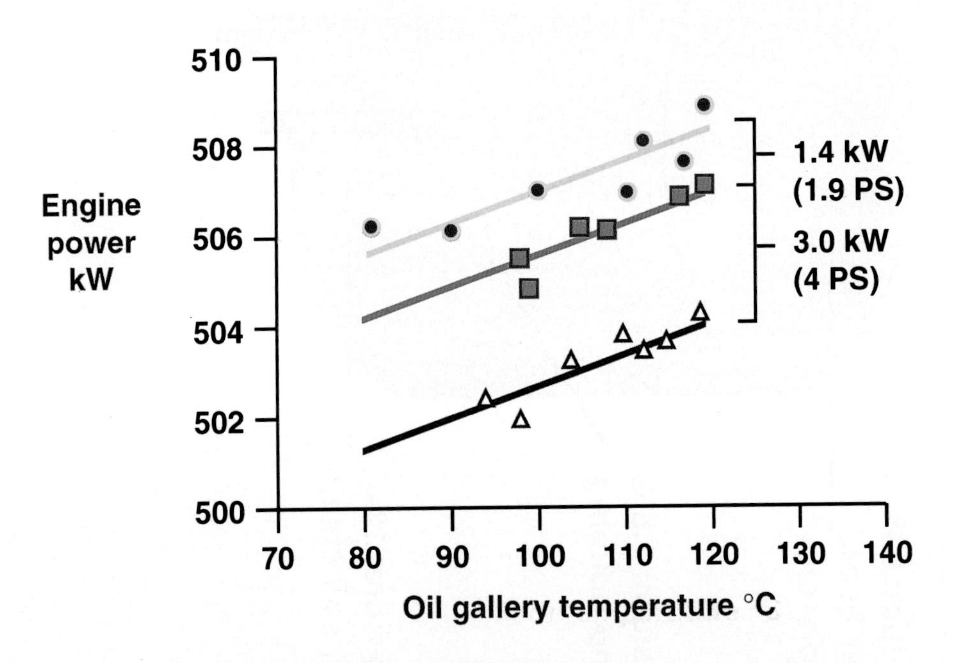

Plate 9 Development of specialist lubrication for F1 engines and the impact on power output (After ref. 11 for the McLaren Honda V12, 1991.) Key: triangles, 20W–50; squares, low viscosity; circles, low viscosity; circles, low viscosity friction, modified.

Power gain through season
70 kW = 95 bhp

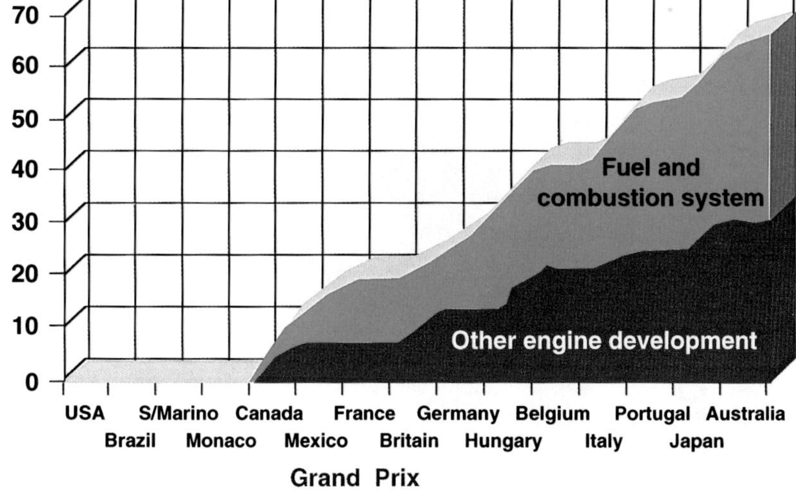

Plate 10 The 'Quality' process of continual improvement in F1 exemplified by the McLaren Honda V12 development in the 1991 season. The graph shows the power gain through the season (70 kW = 95 bhp). Courtesy of Shell Thornton.

Braking Systems

- **Allows much shorter braking distances 170-70 mph in ~ 2 secs**
- **Reduced weight by ~ 15 kg**
- **Low wear rate**
- **No fade**
- **Increased acceleration**
- **Reduced braking effort**

Plate 11 The advantages of carbon/carbon in Formula 1 braking systems.

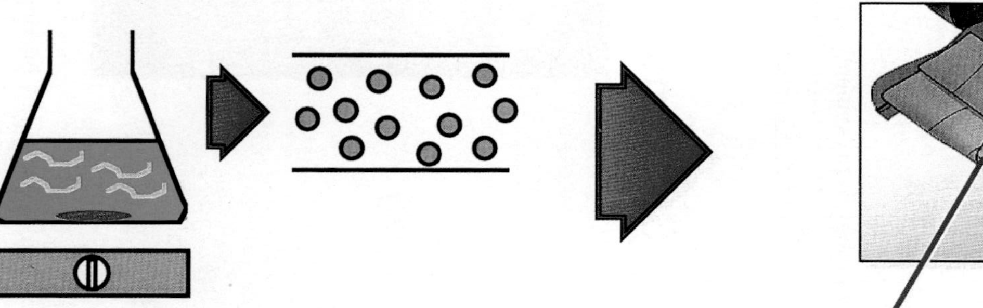

Plate 12 Spatially resolved chemistry to shapes: *the* challenge for the future.

Table 8. Comparison of tyre parameters of different cars

	Formula One car	Sports car	Family saloon
Rigidity of chassis (wheel movement on jacking)	–	4 in	8 in
Relative contact area of tyre	20	3	1
Tyres	Slicks (dry) Treaded (wet) Formula 1 rated	Treaded Zr rated	Treaded Hr rated
Typical tyre lifetime	Qualifying: 10 miles Race: 100 miles	18 000 miles	25 000
Front and rear wings	√	–	–
Steering	Unassisted	Weighted, power assisted	Power assisted

The steel-belted radial tyres used for road-going cars are made of similar materials to F1 tyres (natural and synthetic vulcanized rubbers with steel belting and polyester (or with nylon cord), but the latter are constructed with lower aspect ratios, with variable degrees of carbon black loading (to provide the degree of balance depending on track conditions on 'stickiness and wear') and are made stronger stiffer by incorporation of Kevlar fibres.

It is interesting to note that Faraday himself was one of the first to carry out experiments heating natural rubber with sulphur,[(8)] which is the key to converting the material into a useful resilient material. His experiments in 1825, published in the Royal Institution's proceedings in April 1826, predated Goodyear's key experiments in the USA by some 15 years; even so, a patent on vulcanization was granted in the UK to Hancock 6 weeks before the patent was granted in the USA to Goodyear in 1844, a situation which would not be possible now with virtually instantaneous communication and the ability to search the patent and scientific literature extremely rapidly. In passing it is worthwhile noting that Faraday records that heating natural rubber with sulphur gave a useful preparative route to H_2S. Some of the important tyre parameters of road-going and F1 cars are compared in Table 8. The striking differences are those associated with operating temperatures. The slick F1 tyres work optimally at around 100 °C, hence the need for pre-race electric tyre warmers, the tyres being kept at working temperature during the race by the conversion of mechanical energy to heat associated with the frictional losses which also need to be controlled to avoid excessive heat damage. The composition of the tyres in F1 needs therefore to be optimized for the individual tracks and for the prevailing track conditions. The contact area for a normal family saloon, wearing MOT approved treaded tyres, is roughly equivalent in total to two small hand prints, whilst that for a slick treaded F1 car is some 20 times higher. Cornering forces and traction in general depend on contact area and to control the temperature F1 tyres are typically designed to operate at slightly lower slip angles, as defined in Fig. 9, than for road-going cars.

The other feature evident from Table 8 is that the F1 tyres are designed with maximum grip for a limited mileage. Even with treaded tyres, as during a wet race, the contact area for a F1 car is typically five times that of a family saloon and the tyres can clear an impressive 25 litres of water per second at speeds of up to 180 m.p.h.

The design and composition of tyres for wet and dry races are significantly different and races where one part of the track is dry and other parts wet (very often seen at long race courses like Spa-Francorchamp in Belgium, where it is possible for rain in one part of the track to be accompanied by sunshine in another), pose particular tactical problems for the teams and driver. The nature of the problem is evident from Fig. 10.

Wet (treaded) tyres on a dry track rapidly overheat and lose rubber and, since the contact area is only typically 35 per cent of that for a slick tyre,

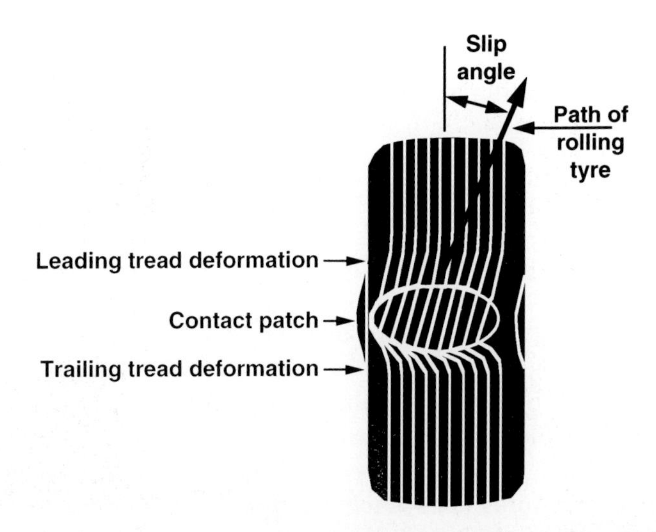

Fig. 9 Slip angles for F1 and road-going tyres. The tyre slip angle is shown as viewed from the road, with successive tread paths depicted on the tyre tread surface.

there is also a massive loss in cornering ability, etc. By contrast, slick tyres have zero grip in the wet (since there is no mechanism to remove the surface water layer) and hence provide neither traction or control. On a drying track, therefore, or if a dry race suddenly turns into a wet one, the teams are faced with some interesting tactical decisions.

The coefficients of friction achievable with short-life 'sticky' tyres are impressive as should be evident from Fig. 11.

I noted earlier that developments in tyre materials and construction have been a major influence in the dramatically increased performance of F1 cars and the narrow-treaded, cross-ply tyres of yesteryear have been replaced by low aspect ratio wider radial ply tyres; this is illustrated in Fig. 12.

The current generation of tyres for road-going cars, owe a lot to motorsport in general and the broad design principles may be stated as follows:

- a longer footprint provides traction and braking
- a wider footprint provides cornering
- a wider footprint provides total grip
- stiff wall constructions and low aspect restrict flexing and deflection under load.

With road-going cars now routinely capable of sustained travel at high speed on the motorways, such developments have contributed very significantly to road safety.

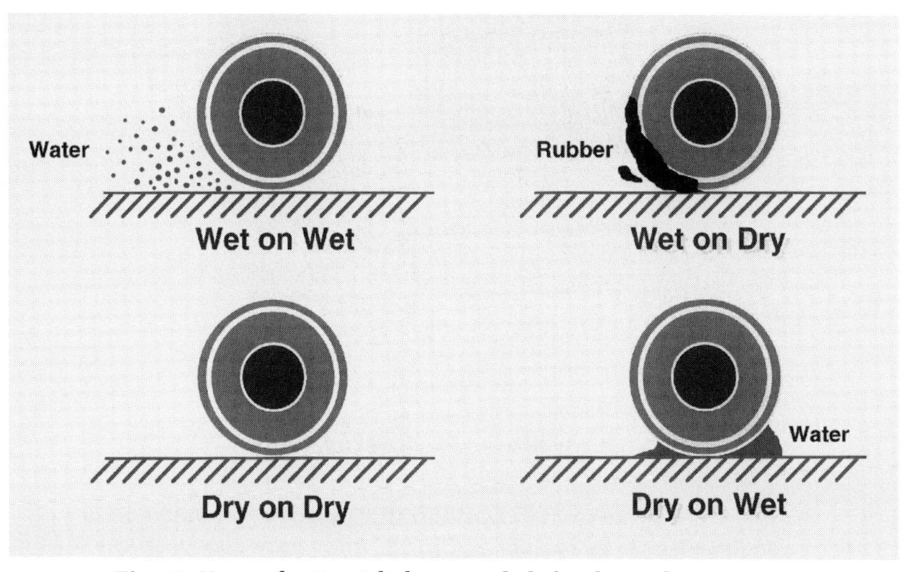

Fig. 10 Tyre selection (slick vs treaded) for dry and wet races.

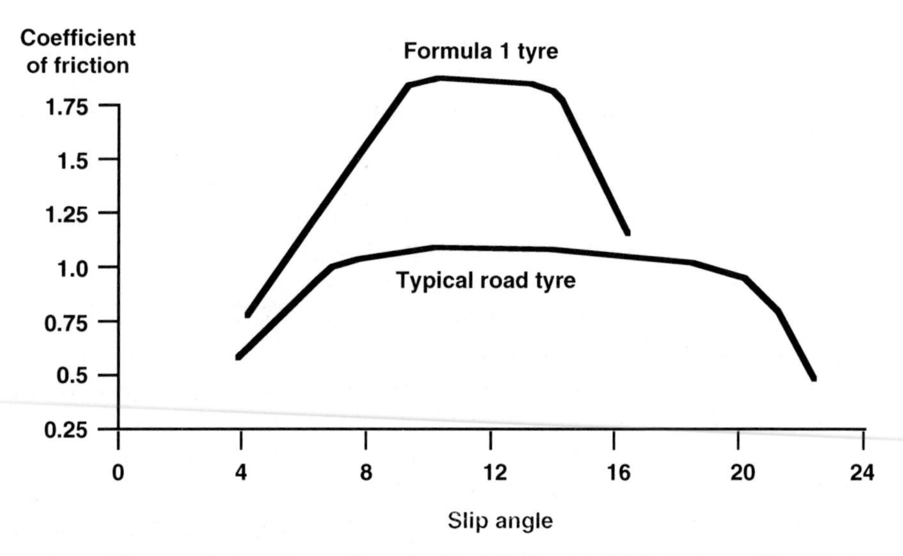

Fig. 11 Comparison of typical coefficients of friction vs slip angle for F1 and road tyres.

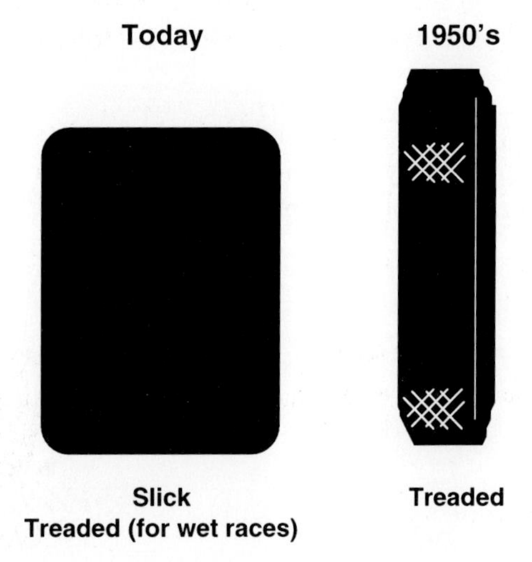

Fig. 12 Comparison of historic treaded and current slick F1 tyres.

Cellular structures

Since a column of air (or gas in general) in a closed cellular structure can have a remarkably high specific stiffness (one only has to think of an inflatable raft or airship to be convinced of this), cellular structures at the nanoscale are quite commonly seen in nature (wood, cork, balsa). At both the macroscopic (honeycomb) level and the nanoscale (foam) level F1 makes extensive use of cellular structures to provide stiffness, lightness, and damage tolerance.

The aluminium honeycomb structures which were used extensively in F1 monocoque structures in the 1970s and 1980s still see use in parts of the F1 'tub' and in the curved surface wings, etc. Honeycombs constructed from Nomex are used in the composite construction of the nose cone where any impact leads to efficient energy dissipation via the crushable deformation of the honeycomb.

One of the most widely used polymers systems in automotive applications in general are polyurethanes (PUs) which can provide an amazing range of closed cell structures with densities down to 0.04 (gm cc^{-1}). Not surprisingly PUs also feature significantly in F1. Figure 13 provides an overview also emphasizing the distinctively different requirements from those for road-going cars where insulation from the environment is often a significant selling point.

The range of frequency-and temperature-dependent mechanical properties that can be engineered into closed cell PU foams is quite remarkable[9] as shown in Fig. 14.

In addition to the use of solid PU materials for the computer numerical control (CNC) machining of full-scale design prototypes of F1 vehicles, the self-skinning properties are useful in providing a comfortable steering wheel, most often fabricated from carbon-fibre composite/alloy, foamed PU shaping, and finally a suede, non-slip covering. PUs are also used to

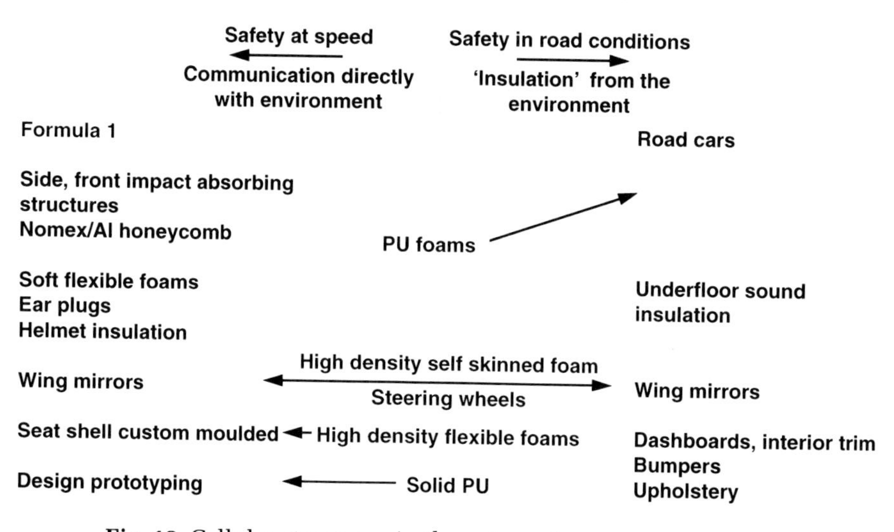

Fig. 13 Cellular structures in the automotive arena.

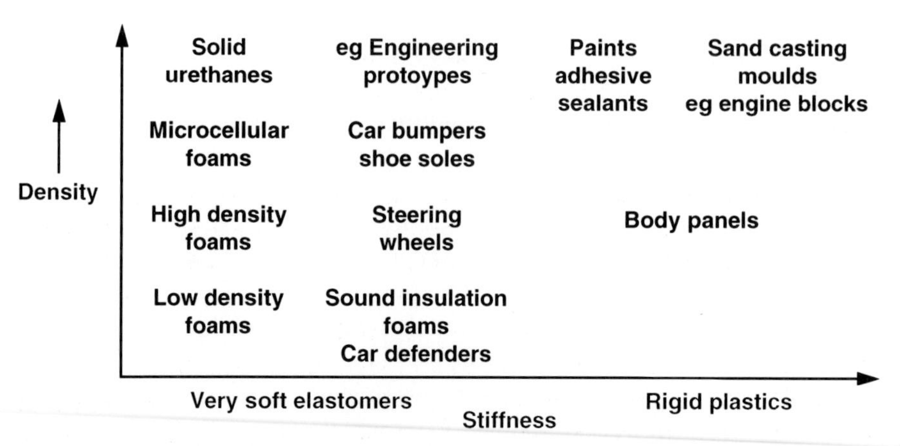

Fig. 14 Schematic illustrating the wide range of properties and applications of polyurethane foams. Densities range from 0.04 to 1.22.

produce a closely contoured master of the driver's shape. This is accomplished by mixing a PU foam and pouring it into a large plastic bag, which is inserted into the driver cockpit whilst the driver supports his weight on his elbows. As the PU foam starts to 'set', the driver lowers himself on to the plastic bag containing the foam, thus providing a negative from which the hand-layered-up carbon fibre/Kevlar composite seat itself is made conforming exactly to the driver's body. As opposed to road-going cars F1 cars have no upholstery—the only cover for the composite shell is again provided by suede.

PU foams are also important in the lightweight carbon fibre/Kevlar composite construction of the driver's helmet.

The engine

The heart of F1 is the engine. Even with the best chassis, it is often the case that the team with a more powerful, tractable engine package can be the overall winner. Even a 10 bhp advantage can make a significant difference and since only one team, Ferrari, manufacture both engine and chassis the teams have to rely on a strategic alliance with a major engine manufacturer (e.g. Williams–Renault and McLaren–Mercedes), or, for the less competitive teams, have to purchase their engines.

The F1 engine is an engineering *tour de force* as is evidenced by the following:

- output > 200 bhp per litre
- extremely rapid response

- rev limits now approaching 16 000 r.p.m.
- pneumatic valve lifters and titanium valve trains
- refined engine management and telemetry

It is interesting to track the development of F1 engines and their outputs over the past 40 years starting with the straight 8 configuration of the 2.4 litre Mercedes of 1954 pushing out 290 bhp through the effective Ford V8 of 1970 when from a 3 litre engine some 510 bhp was produced through to the V10 Renault of the present time when the 3.5 litre engine produces some 800 bhp. Materials have played a significant part in this both in terms of lighter, higher revving engines but also in terms of more efficient lubrication and improved fuels. It should be noted that since power depends on the movement of air and fuel through the engine, and since higher revving engines require a shorter stroke, then piston speeds for F1 engines are significantly higher than for road-going cars. The development of wear-resistant surface engineered coatings for the pistons, etc., is also an area where materials development has had significant impact. Another way of increasing throughput is via turbocharging and, although turbocharged engines are now excluded by the formula, it is interesting to note that the foremost proponents of the technology in F1, Honda, were producing 1200 bhp from 1.5 litre engines in 1987. Since designing and building an F1 engine is so specialized, few companies have attempted to compete. The dominant forces over the years have been Ferrari, Ford (Cosworth), Renault, and Honda.

The typical materials mix of a F1 engine and its ancilliaries is outlined in Plate 7 where it is clear that metals still dominate. There is, however, plenty of scope for the future in moving to lighter, stiffer, carbon-fibre based materials. Surface engineered carbon/carbon pistons could in the longer term have significant advantages as could engine blocks themselves made of carbon-fibre composites. Indeed in other formulas of motorsport such developments are already occurring.

It should be evident from the prominence of the advertising decals that the major oil companies have a large stake in the high profile F1 world. The cutting edge of research in lubrication and in fuel development is one reason for this interest and Shell, Elf, Agip, Mobil, BP, etc., devote considerable research effort to ensure that their sponsored teams remain competitive.

One area of very active research has concerned fuels which with high revving engines and advanced fuel management systems, and the dynamic tuning of the ducting in and out of the cylinders optimally need to show different 'burn' characteristics from standard pump fuels. For example, Plate 8 shows the typical make-up of fuels from the early 1990s compared with the comparable forecourt fuels bought by the average motorist. The specialist hydrocarbon fuels (it should be noted that Indycars are only allowed to use methanol as fuel) responsible for producing significantly higher specific engine outputs in this era typically cost 100 times more than pump fuel and again, to control the technology and competitive advantage from the most advanced teams, 1994 sees current F1 cars using close to standard pump fuel.

Another area of competitive advantage is that involved in polymeric additives for friction and wear control which, with the higher piston speeds, is at its extreme in F1. The particular features required of an efficient oil can be summarized as follows:

- maximum power output without affecting durability
- control of temperature dependence of viscosity
- mechanical shear stability
- wear protection by control of minimum bearing oil thickness
- improved chemical and thermal stability

Plate 9, taken from research done at Shell's Thornton Research Centre[10,11] in collaboration with Honda and McLaren for the 1991 season, shows that even oil developments contribute to small but important increases in horsepower by the use of designed friction modifiers.

The cumulative effective during a season of the 'quality' process of continual improvement both from engine/fuel management design and from the improvements to fuels and oils is dramatically demonstrated from Plate 10 where the Honda V12 with a starting power output probably around 690 bhp improved dramatically to closer to 790 bhp by the end of the season, this being one of the reasons why McLaren won the constructor's and driver's championship.

The future of F1: the materials dimension

I have noted that the FIA, in an attempt to provide more spectator appeal, by reducing the advantage of the technology developers, have banned active suspensions, traction control, special fuels, turbocharging, etc. Paradoxically, this could provide even more impetus for new materials development. I consider one such system.

F1 currently is forced to use the same carbon/carbon composite brake materials as one would find on a Boeing, Airbus or indeed military aircraft such as the Stealth bomber, itself being constructed largely of the same type of carbon-fibre composites as used for the F1 monocoque. The braking characteristics of an F1 car are not analogous to those of an aeroplane which really only needs braking of carbon/carbon capabilities for an aborted full-load take-off, or for landing. By contrast an F1 driver is continually on and off his brakes, so the friction and wear are unlikely to be the same as for a jumbo jet. Because of the cost, however (carbon/carbon can cost up to £1000 kg), F1 typically uses the centre 'offcuts' from aerospace. This in turn means that in order to maintain optimum temperature as a compromise between frictional properties and oxidative wear, complex track dependent ducting to the brake disks need to be employed. The desirable features of carbon/carbon brakes in F1 are illustrated in Plate 11. Figure 15 gives a comparable analysis for the other areas of great competitive advantage in carbon/carbon application in F1, namely clutch design. This suggests two obvious materials-driven goals, firstly intensified cheaper generic routes to carbon/carbon so that materials optimum for F1 can be developed independent of aerospace; and secondly surface engineered systems that give the required tribological properties without oxidative wear such that the ducting can be dispensed with, simplifying design of braking systems. Using **Transform** concepts this is now being addressed using novel reactive hot isostatic processing technology' being developed by Surface Transforms Ltd.[12]

An F1 car with traction control, advanced ABS, and active suspension is an example of a rudimentary 'smart materials' system, in that it can sense and respond to its environment.[13] Although this is currently achieved by integrating sensors, accelerometers, and actuators of macroscopic design into the materials system, long term we can see that materials will need to develop closer analogy with nature where structures

- **Rapid disengagement time**
- **Rapid recovery from clutch fluid or oil**
  (C/C conventional clutch material)
- **Extended life**
- **Technology diffusion into industrial sector**
  - **Heavy plant and machinery**
  - **Exotic sports cars**

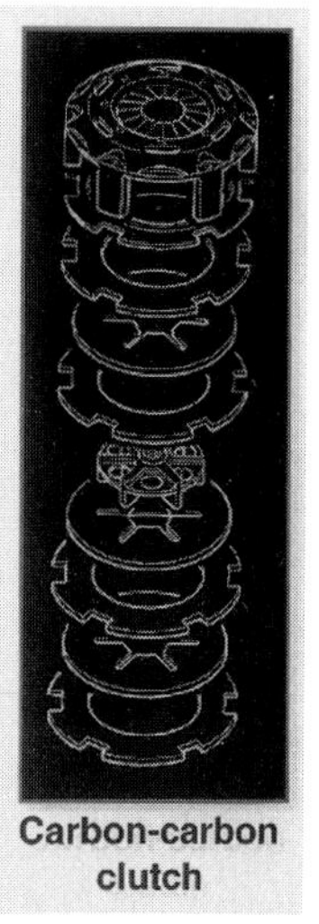

Fig. 15 Advantages of carbon/carbon in F1 clutch design.

optimum for function that are active and self-repairing are effectively designed at the molecular level. This is illustrated below:

- 'Smart materials' systems
  Active
  Responsive and interpretive of their surroundings
  Self-repairable long term
- Long-term requirement is the ability to do spatially resolved chemistry direct to functional shapes

A key issue is how to achieve synthesis and assembly directly, rather than as now in the chemical/materials world, where essentially all of the synthesis is in the unshaped form, which then goes through a preshaping and ultimately a shaping process with the properties of the final shape also hinging critically on the long-and short-range order and changes therein arising during the shaping process. A key challenge for the future, therefore, are the issues of synthesis, shaping, and assembly, as an integrated process, and F1 with its rapid time-scales might well be a suitable vehicle to demonstrate some of the emerging science and technology to address such issues, particularly via 'Transform engineering'. Plate 12 for example, shows a schematic of a 'high IQ' self-adapting, self-repairing wing assembly.

Although we may anticipate the '**smart**' technology will be all-pervasive by the year 2020, the likely scenario is a rapid introduction not only to specialist fields like F1 but also to commodity consumer sports goods like smart golf clubs and tennis racquets. For example, one could envisage a 'smart' golf club. This might have two dials relating to playing style, etc., the essence of the smartness being a head which sensed before impact whether you were a slicer or a hooker and that dynamically adjusted the face and the 'sweet spot' (centre of gravity) before impact to give you a perfect shot. This would allow a competitive game between a novice and an expert without the crude resort to handicapping.

It is clear from the current F1 experience that a small degree of smartness gives exceptional competitive advantage and with future developments the prospect for continued and seminal impact on F1 by the materials world is an exciting one.

Acknowledgements

I would like to thank the following for their help in pulling together the lecture and exhibition materials for the Discourse: David Williams and Brian O'Rourke (Williams Grand Prix Engineering, Didcot), Fred Pfister (Dupont, Geneva), Ian Galliard and Toshiyuki Ozuma (Shell Research,

Thornton), Roy McCarthy (Dowty Aerospace Propellors, Gloucester), David Clarke (Dunlop Motorsport, Birmingham), Colin Hill (Rk Carbon, Altrincham), Tony Cunningham and Geert Bleys (ICI PU, Everberg), Bryson Gore (Royal Institution, London), Terry Walker (ASALS, Indiana), Ron Marl (SERC RUSTI, Daresbury), Nick Tiffin (ICI Fiberite, Warwick), and John Evans (TIA Graphics, Manchester).

References

1. Cimarosti, A. (1990). *The complete history of Grand Prix motor racing*. Motor Racing Publications, Croydon.
2. Henry, A. (ed.) (1992). *Autocourse 1991–92 yearbook* Hazelton Publishing, Richmond
3. Henry, A. (ed.) (1994). *Autocourse 1993–94 yearbook*. Hazelton Publishing, Richmond .
4. Shaw, J. (ed.) (1994). *Autocourse—Indycar yearbook*. Hazelton Publishing, Richmond.
5. Gordon, J. E. (1988). *The science of structures and materials*. Scientific American Library, New York.
6. Cogswell, F. N. (1992). *Thermoplastic aromatic composites*. Butterworth-Heinemann, Oxford.
7. Clark, D. T. (1993). Macromol. Chem. Macromol. Symp., **75**, 1–34.
8. Faraday, M. (1826). *Q.J. Sci.*, **21**, 19–28.
9. Woods, F. (1990). *The ICI polyurethanes book*, 2nd edn. Wiley, New York.
10. Glover, R. and Galliard, I. (1993). *Fuelling success in Formula One*. Shell International, London.
11. Miller, S. and Pinchin, R. (1990). *Engine lubrication for the Formula One world champions*. Shell International, London.
12. Clark, D. T. (1993). *Advanced Materials*, **5**, 502–7.
13. Clark, D. T. (1994). *Materials World*, **2** 144–5.

D. T. CLARK

Born 1939, obtained a Ph.D. from the University of Sheffield. Was a Fulbright Fellow at Caltech before becoming Lecturer in Chemistry at Durham University, then, successively, Reader, Professor, and Department Head. From 1983–93 he was successively Research Director for ICI's New Science Group and Lab Director and Research Manager for ICI's Wilton Materials Research Centre. During this phase of his career he became significantly involved in the ICI–Williams interaction. In 1993, he became Scientific Director of SERC Daresbury's Research Unit in Surfaces, Transforms and Interfaces, and is developing new market applications for transforms via a high-technology company, Surface Transforms, of which he is Chairman. A keen follower of Formula 1, he has driven cars around many of the F1 tracks in the UK and Europe.

Hello sunshine

MARY ARCHER

It never set on the British empire. Autumn is its close-bosom friend, but Mrs Ogmore-Pritchard required it to wipe its feet before it came in under Milk Wood. You can buy a cheap version of it for 20 pence. It led Icarus to his sticky end, although it has nothin' to do 'cept roll around heaven all day. There's no new thing under it, but we hope to have our place in it, and at its going down we will remember them.

It evolved from a glowing gas cloud into an unremarkable yellow star. It is currently in sedate middle age, with about five billion years to go before it exhausts its nuclear fuel and blows up into a red giant, devouring the Earth in flames. Finally, at the clockwise end of its evolutionary spiral, it will become a dead white dwarf. Round it move the planets in their immutable orbits, an arrangement that so far removes Man from the centre of things as gravely to have displeased the Inquisition in Rome before which, nearly four centuries ago, Galileo upheld the heliocentric astronomy of Copernicus and Aristarchus.

The '*it*' of which I speak is, of course, the Sun, the Sun which has provided the warmth and light for the evolution and sustenance of life on Earth. Not that this magnificent heavenly body has historically been regarded as a mere it; for Louis XIV of France, the Sun was male and symbolized the potency, power, and glory of the Sun King himself. But for the Bedouin of Arabia, who had altogether too much of it, the Sun was a destructive old woman, who forced the handsome Moon to sleep with her once a month, and so exhausted him that he needed another month to recover.[1]

The Earth's fuel and energy resources

We are the merest upstarts in the solar system, as we may see from Carl Sagan's brief history of the world compressed into one year (Fig. 1). On

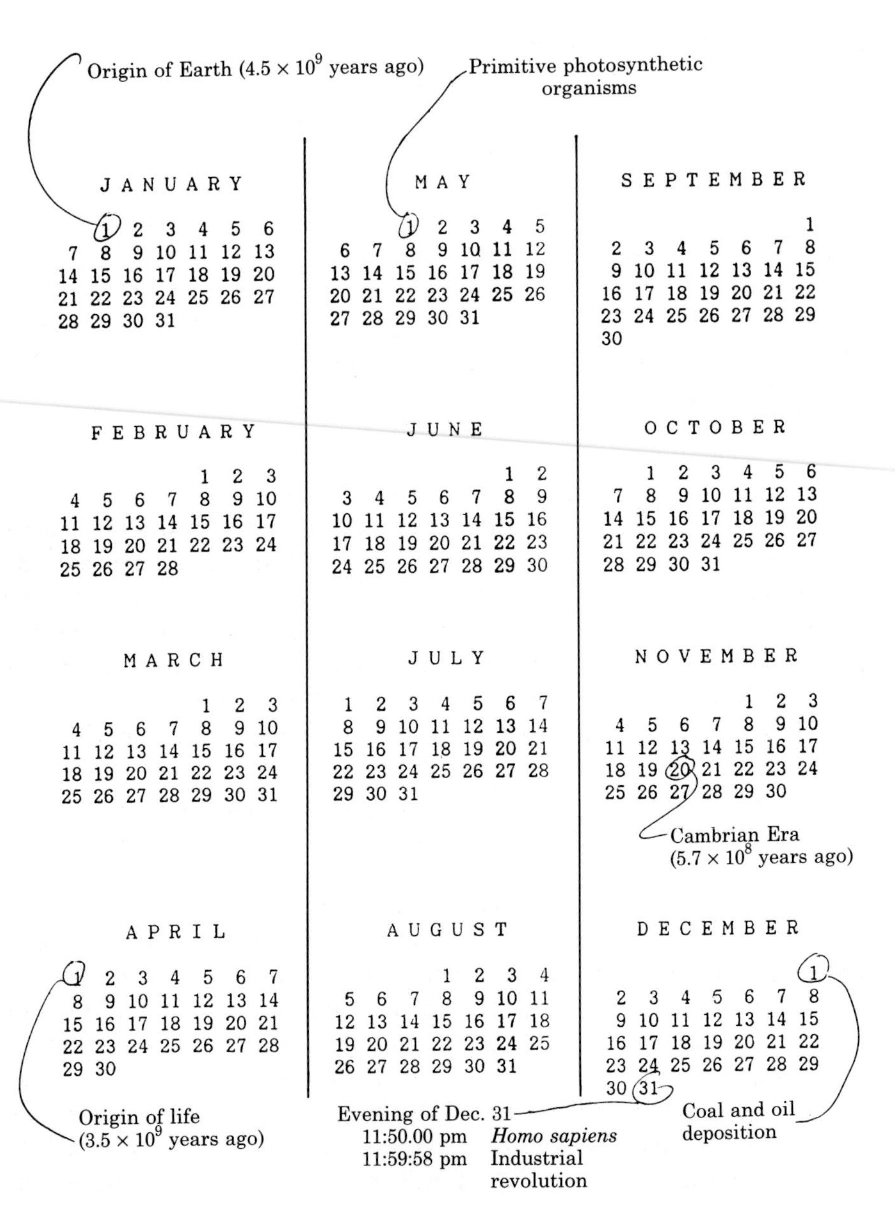

Fig. 1 The history of the Earth compressed into one calendar year (adapted from C. Sagan (1977), *The Dragons of Eden*, Random House, New York.

this time scale, 1 January represents the formation of the Earth by the condensation of interstellar matter some four and a half billion years ago, and it is now midnight on New Year's Eve. In this time frame, the Earth awoke from its long Precambrian slumber only in mid-November. The coal, oil, and gas, which at present supply about three-quarters of the world's total energy requirements, were laid down in early December. The rather optimistically named *Homo sapiens* appeared only today—at 10 minutes to midnight. The Industrial Revolution began two seconds ago, powered by coal and latterly by oil and gas, and in those two seconds we have consumed perhaps one quarter of our total stocks of fossil fuels. Within the next few seconds—within the next few centuries in real time—we will in all likelihood devour the rest.

If that is so, we are midnight's children, striking one short match in the middle of a long dark night. When the hydrocarbon match burns low, assuming that we have not by then discovered some new exploitable force of nature, we will have to turn to some combination of nuclear and renewable energy, or maybe return to a combination of the two, for the Sun is a nuclear fusion reactor. At the enormous temperatures and pressures of the inner Sun, four hydrogen nuclei fuse to form one helium nucleus. Some matter is destroyed in this fusion process, and when matter is destroyed, energy is created in accordance with Einstein's seminal equation $E = mc^2$ where E is the energy created, m the mass destroyed, and c the speed of light. It is impossible to travel faster than the speed of light and, as Woody Allen has pointed out, it is also undesirable because one's hat keeps blowing off.

The Sun is losing mass at the staggering rate of 4.5 billion kg sec^{-1}, outpouring about 10^{26} W of radiant energy into space. Our eyes are so good at accommodating to different light levels that it is difficult to believe that full sunlight on Earth is some half a million times brighter than full moonlight. Not that this impressed Sydney Smith, writing to satirize the dismissive style of his friend, Francis Jeffrey, 25 February 1807: '*Damn the solar system! bad light—planets too distant—pestered with comets—feeble contrivance;—could make a better with great ease.*'[(2)]

But we do not need a better, as we may see from our global energy account (Fig. 2). The scale is logarithmic, so that each rung represents 10 times more energy than the rung below. On the left is shown our energy capital—the Earth's estimated remaining resources of coal, gas, and oil, and of fissile uranium, used today in non-breeder reactors, tomorrow maybe in breeder mode. Also on the left is shown our global energy expenditure, that is, the world's energy demand in the years 1970 and 1990, plus a prediction for the year 2030. Clearly we are living off capital, dwindling fossil fuel capital. And it would not be wise to bank on the

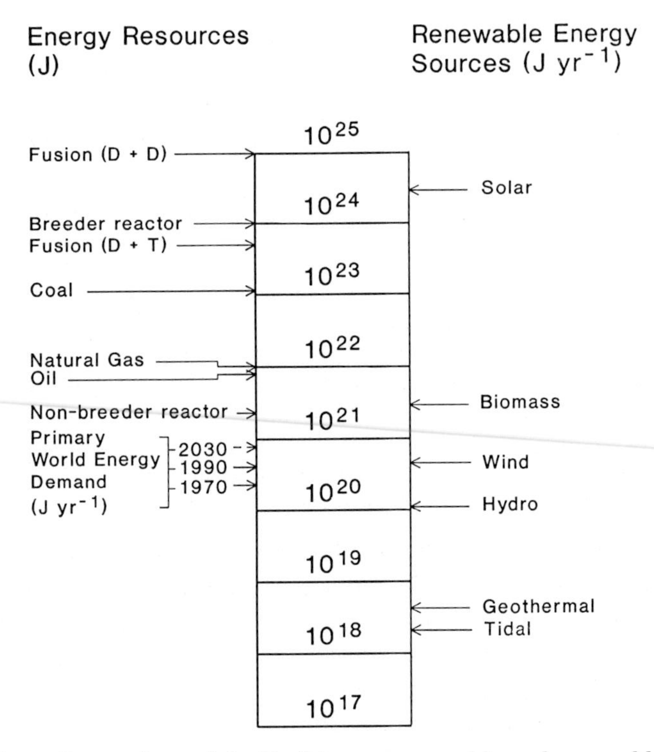

Fig. 2 Comparison of the Earth's non-renewable and renewable energy resources on a logarithmic scale.

capital represented by the fusion fuels deuterium and tritium, because it is by no means certain that commercial fusion power can be achieved.

To be prudent, we should therefore look for another source of income, and there it is, on the right-hand side of the ladder, the annual energy available from our renewable energy resources. The great fusion reactor in the sky, the Sun, is by far the most abundant of these. Indeed, the energy from the Sun falling on the Earth in one hour is equivalent to the world's energy demand in one year. The other renewables—wind, biomass, waves, and so on—are mostly driven by the Sun. As Harold Heywood, a great British pioneer of solar energy utilization, observed in his Discourse of 1957, the subsidiary renewables represent the concentration of solar energy in space, just as fossil fuels represent its concentration in time.

Characteristics of solar energy

Worship of the Sun goes back to the cult of the Sun god, Ra, at Heliopolis and beyond. As to use of the Sun, the Greeks understood and practised the principles of solar architecture, although the account of Archimedes setting fire to the Roman fleet off Syracuse by reflecting sunlight off the polished shields of the Greek soldiers is probably apocryphal. But the chemist's burning glass is real enough, and it was with such a glass that Joseph Priestley isolated 'dephlogisticated air'—oxygen—in August 1774, by focusing the Sun's rays onto red oxide of mercury, and confining a mouse under a bell jar of the evolved gas:

> Having procured a lens of twelve inches diameter and twenty inches focal distance, I proceeded ... to examine what kind of air a great variety of substances would yield. I presently found that, by means of this lens, air was expelled from Mercurius Calcinatus very readily. But what surprised me more than I can express was that a candle burned in this new air with a remarkably vigorous flame, and my mouse lived in it for half an hour ... I was utterly at a loss how to account for it.

The mouse did not actually die after the half hour—it was revived by putting it in front of a warm fire, after which it went back into the bell jar for another three-quarters of an hour, once more emerging in reasonable shape.

There are various characteristics of solar energy of which we should be aware in designing solar converters. The first is its relatively low flux density. When the Sun is overhead on a clear day, each square metre of the Earth's surface receives up to 1 kW of solar power. But the Sun is not always out, let alone overhead, so the average power is lower than this, ranging from about 300 W m^{-2} in hot spots like Kuwait to about 100 W m^{-2} in rather less sunny locations such as Manchester (Fig 3). The amount of solar energy received varies seasonally, particularly at high latitudes. The summertime maximum is less marked on a vertical surface than on a horizontal surface because the vertical surface sees the low winter Sun better. Fixed, flat plate collectors are usually tilted at or near the angle of latitude to maximize their annual insolation, which comes partly straight from the Sun, as beam radiation, and partly out of the sky, as diffuse radiation scattered out of the beam by molecules or particles in the atmosphere.

Concentrating collectors, like the celebrated solar furnace of Odeillo in the French Pyrenées, use lenses or mirrors to focus sunlight, and work only with the beam component, although there are non-imaging collectors of various types that will concentrate diffuse light. For example, a high refractive index sheet of plastic which contains a fluorescent dye will trap diffuse light incident on the plane face, channelling it out through the narrow edges.

The last notable characteristic of sunlight is that it is white, containing all the colours of the rainbow and more, on into the ultraviolet and infra-

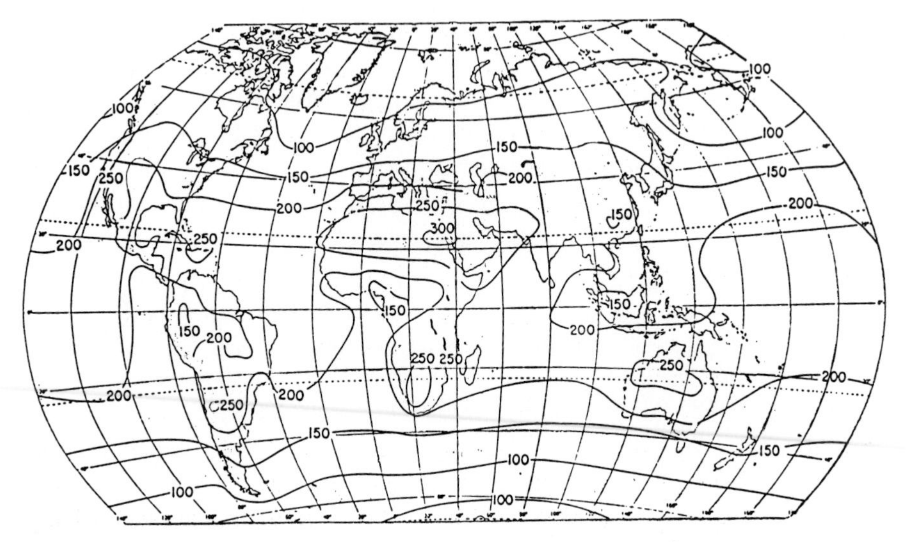

Fig. 3 World distribution of solar irradiance (W m^{-2}) on a horizontal sea-level surface, averaged over a 24 h day.

red. The Sun's surface temperature is 6000 K, and the spectrum of the radiant energy it emits is much like that of a white-hot solid body, except for the tell-tale dark Fraunhofer lines which reveal the elements present in the outer solar atmosphere. A solar converter must absorb and use a good portion of this visible spectrum if it is to be efficient.

Thermal conversion of solar energy

There are two classes of solar converter: thermal and direct. Thermal converters deliver heat as such or use a heat engine to deliver mechanical power or electricity. Direct converters produce electrical power or chemical fuels without the intervention of a thermal stage. Let us consider a model solar community—'Solarville'—with examples of both converter types. In the Industrial Park, there is a solar thermal mill powered by a 60 W light bulb (which simulates the Sun), whose workings we shall shortly examine.

Solar thermal applications go back to the nineteenth century. The solar thermal locomotive designed by the Punch cartoonist Emett has yet to find a backer—though the 1901 vintage 15 horse power solar pump that irrigated the Pasadena Ostrich Farm surely comes close to it—and Augustin Mouchot's Sun Machine of 1878, devised because of the acute shortage of coal in nineteenth century France, finds a modern echo in the dish collectors at Sandia National Laboratory in Albuquerque. Meanwhile, a dispassionate Martian visitor to the site of Eurelios in Adrano might well conclude that the cult of the Sun god is alive and well in twentieth-century Sicily. In fact, Eurelios is the world's first 1 MW solar thermal power tower. The field of mirrors is controlled to reflect and focus the solar beam onto the central receiver, where the temperature can approach 3000 K, and superheated steam is raised to drive an electric generator. They do it bigger and better in California, of course. The power tower Solar One at Barstow supplied 10 MW, but even that is dwarfed by the 354 MW of solar thermal power installed by the Israeli company Luz in the Mojave Desert, which provides two per cent of Southern California Edison's grid capacity.

Let us now see how we can turn some of the heat in a cup of coffee into mechanical work to power the second mill on the Solarville Industrial Park. These mills contain miniature versions of the engine devised by the Scottish cleric Robert Stirling in 1816, and we can run a baby Stirling engine between the temperatures of hot coffee and iced whisky and soda. To run, the engine requires a hot source (in this case a cafetière) and a cold sink, in this case a tumbler of iced whisky and soda—and away goes the engine, with modest efficiency admittedly, but it's simple and reliable. To make it solar-powered, we would have only have to provide the heat by a solar water heater.

Passive solar design

The Sun's heat can be used as such to provide water and space heating for buildings, which may declare their technology boldly or may have less obvious architectural features that maximize useful solar gain and minimize the house's conventional energy demand. We can quantify the difference that passive solar design can make using the National Home Energy Rating (NHER) scale. This goes from 0 to 10, and the higher a house is on the scale, the lower are its energy running costs. The energy rating is calculated from the characteristics of a house—its room and window dimensions, wall construction and roof insulation, fuel and heating type, orientation and degree of shelter, and so on. All this information is fed into the computer, and the NHER software computes the rating.

Using the expertise of Peter Rickaby, of energy consultants Rickaby Thompson, the energy ratings of two similar houses were compared (Table 1). One house, in Newport Pagnell, has no solar features. Its South façade is largely taken up by garage and entrance while most of the living rooms and conservatory are on the North side. This house rates a respectable 8.0 on the NHER scale and has annual energy running costs of £777.

Table 1. NHER ratings of a modern non-solar house and a modern passive solar house

	Non-solar house		Passive solar house	
NHER	8.0		9.1	
Building Energy Performance Index	144		150	
Standard Assessment Procedure	78		88	
CO_2 emissions (tonnes/year)	8.4		5.3	
Analysis of fuel use and costs	*GJ/year*	*£/year*	*GJ/year*	*£/year*
(i) By application				
Primary heating	44.0	192	20.7	90
Secondary heating	9.6	39	5.2	23
Water heating (main fuel)	25.9	113	16.9	74
Cooking (main fuel)	2.9	13	5.4	23
(secondary fuel)	1.7	37	–	–
Lighting and appliances	11.9	267	12.2	261
(off-peak component)	3.0	22	–	–
Standing charges	–	93	–	80
TOTAL	**98.9**	**777**	**60.3**	**551**
Maintenance	–	24		24
(ii) By fuel type				
Gas (mains)	72.8	355	48.1	247
Housecoal/Pearls	9.6	39	–	–
Economy 7 (on-peak)	13.5	346	–	–
(off-peak)	3.0	36	–	–
Domestic tariff (on-peak)	–	–	12.2	303

The bar chart shows where the heat goes—through the walls and windows, wasted in heating appliances, and so on.

The other house has passive solar features; it is in Milton Keynes. Here all the habitable rooms are on the South side, which also boasts a semi-integral conservatory. As a consequence, the South façade has much more glazing than the North side, which houses the bathrooms and passage-ways. The house also has excellent heating controls, needed to take full advantage of solar gain. This house rates an NHER of 9.1, an excellent score, over an integer better than the other house, and its energy running cost is only £551 per annum (Fig. 4b, Table 1). The heating system does not have to work so hard because the house gets more of its heat from the Sun.

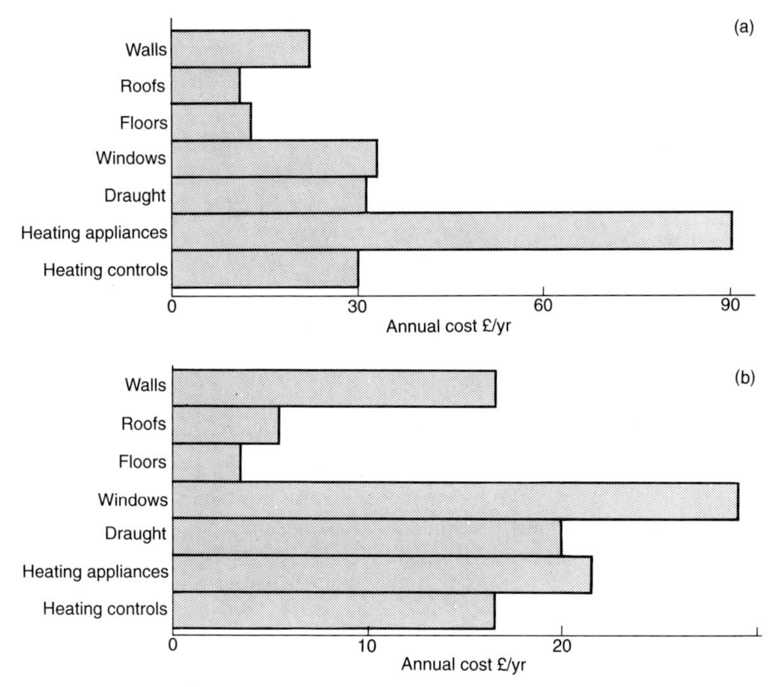

Fig. 4 Breakdown of costs of heating a modern non-solar house (a), and a modern passive solar house, (b).

Photosynthesis

So much for solar thermal conversion. Turning to direct solar converters, these use sunlight not as heat but as a stream of high energy photons, capable of driving processes that store energy. Like many of the ideas I shall mention, this one is not new. The distinguished Armenian photochemist, Giacomo Ciamician, who worked at the University of Bologna at the turn of the century, using the Sun as his light source, predicted the possibility of chemical storage of solar energy thus:

> On the arid lands there will spring up industrial colonies without smoke ... forests of glass tubes will extend over the plains and glass buildings will rise everywhere; inside of these will take place the photochemical processes that hitherto have been the guarded secret of the plants, but that will have been mastered by human industry which will know how to make them bear even more abundant fruit than nature, for nature is not in a hurry and mankind is.[3]

The 'guarded secret of the plants' is the mechanism of natural photosynthesis, the process by which atmospheric carbon dioxide and water are fixed under the driving force of sunlight to form carbohydrate; the myriad genera of green plants are Nature's advanced direct solar converters. They have evolved from primitive photosynthetic organisms which appeared as early as 1 May in the time-compressed history of the world. All photosynthetic organisms contain organic pigments which absorb sunlight and initiate the process of photosynthesis. Chief among these are the chlorophylls, of various shades of green, the carotenoids, whose orange or yellow colours are seen in autumn leaves in which the chlorophyll has decomposed, and the phycobilin pigments such as phycocyanin.

If these pigments are illuminated with ultraviolet light, chlorophyll fluoresces red and phycocyanin orange. Cartenoids do not fluoresce, and for a good reason—they are there to protect the plant against excessive light levels. Fluorescence comes about because the absorption of a photon—a quantum of light—promotes an electron from its ground band in the pigment molecule to an upper band, leaving a hole in the ground band. This process stores energy in the hole–electron pair, equal to the energy difference between the two bands. The hole–electron pair is not long-lived and it can recombine in three ways. The electron can return directly to the ground state, giving up its stored energy as fluorescence, which is what is happening in the chlorophyll and phycocyanin. Or the stored energy can be given up as heat as the electron returns to the ground state, as in the carotenoid. Either way the energy of the hole–electron pair is wasted. But if the system is so designed, the electron can be made to do work as it returns to the ground state. We can represent this as mechanical work by making the electron of our mechanical model travel back to ground level in a basket, raising a small weight as it goes.

There is a school of thought that holds that, since aeroplanes do not flap their wings like birds, successful man-made direct converters need be nothing like a green plant. But it turns out that the 'guarded secret' of the plants, as revealed by modern structural analysis, has many valuable lessons for us. A leaf is a highly structured system at the macroscopic, microscopic, and atomic levels. All the important molecules are held in very specific locations and orientations with respect to one another, as we can see from the structure of the reaction centre of the photosynthetic bacterium *Rhodopseudomonas viridis* (Fig. 5), which was crystallized and analyzed in the mid-1980s by Hüber and Michel at the Max-Planck-Institute in Martinsried, in a bravura piece of work that won them the 1988 Nobel prize for Chemistry.

The heart of the reaction centre is a special pair of bacteriochlorophyll (*BC*) molecules. When these receive a photon of sunlight, an electron is excited and moves very rapidly via the bacteriopheophytin (*BP*) molecules

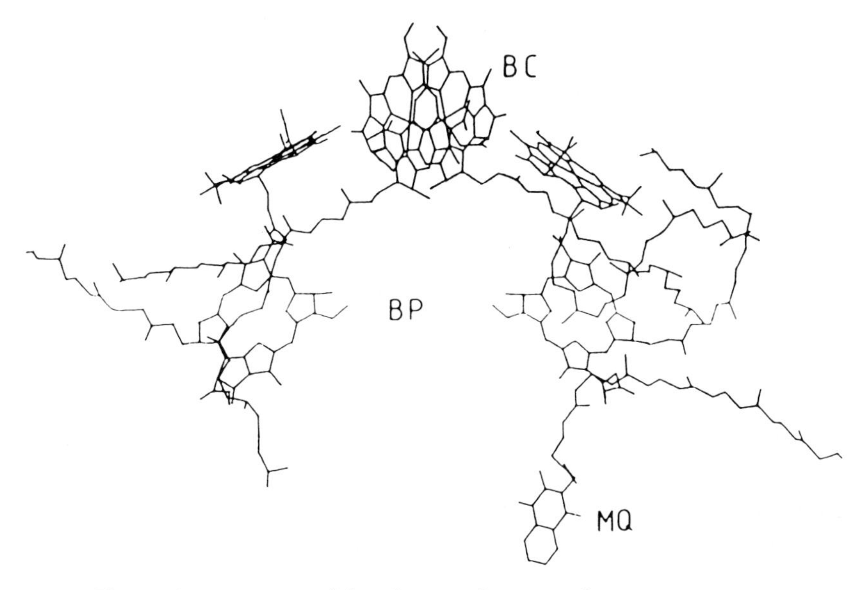

Fig. 5 Arrangement of the chromophores in the reaction centre of *Rhodospirillum viridis*. From J. Deisenhofer, O. Epp, K. Miki, R. Huber, and H. Michel (1984) *J. Mol. Biol.*, **80**, 385.

to the quinone molecule (*MQ*) on the lower right-hand side. Then a sequence of events turns the energy of the hole–electron pair—the electron on the quinone and the hole left behind on the chlorophyll—into chemical energy stored in carbohydrate. The point I want to make here is simply this: the structure of the photosynthetic reaction centre is the key to its function. If we blended it all up like the pigments in solution, the resulting 'leaf soup' would not photosynthesize any carbohydrate whatsoever.

Now that the structure of the natural photosynthesis reaction centre is known, chemists are addressing themselves to the formidable challenge of making a simple working version in the laboratory. In such synthetic assemblies, it is now possible to observe electron transfer occurring as it does in natural photosynthetic organisms, and the long-term hope is that one day chemists will be able to synthesize systems capable of fixing carbon dioxide in a test tube. Maybe only God can make a tree but, as Lord Porter observed, Man may retain the more humble ambition of making a leaf.

Biomass and energy crops

What about growing crops for their energy content rather than for food? The difficulties of making biomass cost-effective emerged very clearly when Gulliver encountered on his travels in Laputa a Scientist, ‘a Man . . . of meagre Aspect with Sooty Hands and Face’:

> He had been Eight Years upon a Project for extracting Sun-Beams out of Cucumbers which were to be put into Vials hermetically sealed and let out to warm the Air in raw inclement Summers. He told me that he did not doubt in Eight Years more, that he should be able to supply the Governor’s Gardens with Sunshine at a reasonable Rate; but he complained that his Stock was low, and intreated me to give him something as an Encouragement to Ingenuity especially as this had been a very dear Season for Cucumbers.

Jonathan Swift notwithstanding, the economic attractions of energy crops have been much increased by the policy of agricultural set-aside. By the turn of the century, there could be 20 million hectares of surplus agricultural land within the European Union as a whole, an area equivalent to England, Wales, and a large part of Scotland. Farmers cannot grow food on this land, but they could grow energy crops. There are several candidate crops, among which it has to be said that cucumbers, with their 98 per cent water content, are not front runners. *Miscanthus*, a perennial woody grass, grows very rapidly and can be harvested by familiar means. Rape is now grown as a subsidized energy crop on over quarter of a million hectares in Europe, the oil from its seed being esterified and marketed as biodiesel; in Austria, this now accounts for about five per cent of the total diesel fuel market. But wood, although less familiar to the farmer than the forester, looks like the best bet for the UK. Willow and poplar both root readily from simple cuttings, which to be grown as an energy crop are planted at high density. These grow rapidly and can be harvested by coppicing, that is, by cutting down the long stems and leaving the stools to sprout again, as they readily do. The energy content of wood is only about half that of coal, and it is generally preferable to use such a low density fuel near where it is grown. In Sweden, for example, there are some thousands of hectares under willow for use in district heating schemes.

Photovoltaic cells

Between the sunlight incident on the green leaf and the energy stored in the crop lies a long and complex biosynthetic pathway. So it is no wonder that the net energy storage efficiency represented by a bundle of wood is very modest, perhaps one half to one per cent. Anyway, efficiency is not really the right criterion for an energy crop—price, sustainability, and carbon dioxide abatement are all more important considerations. In terms of sheer efficiency, Man overtook Nature forty years ago with the silicon photovoltaic (PV) cell. PV cells, or solar cells as they are sometimes known, are direct converters that produce electric power when they are illuminated. Solar cells are made of semiconductors, and if for example, an old de la Rue set of semiconductors is illuminated with ultraviolet light, they fluoresce, showing that electrons are excited when the semiconductors absorb light, just as they are in photosynthetic pigments. Of course, a solar cell is altogether a much simpler structure than a green leaf, but it is like a green leaf in that one side is designed for looking at the Sun. Just under the top surface, there is a junction between two different semiconductors, in this example *n*-type silicon and *p*-type silicon (Fig. 6). In this junction there is a built-in electric field, which effectively tilts the ground and upper bands in the interface, causing the excited electron in the upper band to flow one way and the hole in the ground band the other. In other words, negative charge flows one way across the junction and positive charge the other. Thus the output of a solar cell is electrical work.

Solarville has quite a collection of PV-powered items (and I would like to thank Bernard McNelis, Bob Hill, Phil Wolfe, and Michael Penney for letting Bryson Gore and me play with their toys in the Discourse on which this article is based). The house has a PV roof which provides electric power for the internal lights and appliances; the residents have a PV-

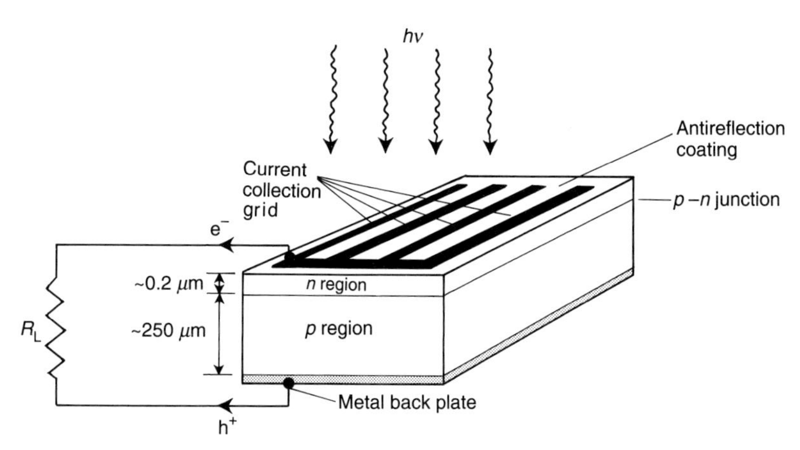

Fig. 6 Simplified structure of a silicon *p*–*n* homojunction cell. On illumination, photogenerated electrons (e^-) flow out of the *n* side of the junction through the external electric load R_L, where they do electrical work as they recombine with holes (h^+) flowing from the *p* side.

driven radio. The Solarville train is powered by a PV module in the centre of the track. PV cells produce power only when they are illuminated, so when the Sun goes behind a cloud, the train stops, which makes for an unreliable service. Appliances that must run in the dark, like the train driver's lantern, therefore have storage batteries than can be recharged by exposure of their PV modules to the Sun.

Modern silicon cells are mostly made of the semiconductor silicon, but cells based on selenium go back to the 1870s. Early photographic exposure meters contained selenium cells, as did the 1960s vintage Agilux camera, which did not require a battery. Other early devices contained copper oxide cells. But the efficiency of these devices was well below one per cent. Then in 1946, Russell Ohl, a metallurgist at the Bell Telephone Laboratory in New Jersey, made a very interesting observation:

> I had made up a considerable number of melts of pure silicon for the purpose of making specific resistance measurements ... I was making such a measurement when I noticed, while viewing on an oscilloscope the wave shape of the 60-cycle current flowing through the rod, that the current in one direction was affected by light from an ordinary 40-watt desk lamp.

In crystallizing the silicon, Ohl had unintentionally made a *p–n* junction, because unsuspected impurities in the melt had stratified in different zones in the solid. Many great scientific discoveries have been made by accident, of course, but only by those who recognize the significance of the surprising, as did Ohl:

> This entirely unexpected phenomenon was recognized as of possibly great importance in the art of light-sensitive devices and further study was undertaken forthwith. The outcome of such study is that improved light-sensitive devices and particularly photovoltaic cells of high sensitivity and great stability have been made available.'

Modern crystalline silicon cells are made from boules of single crystal silicon or cast ingots of multicrystalline silicon. These are sawn into wafers and processed into cells, but the wafers are relatively thick and mechanically fragile. A different form of silicon—amorphous or disordered silicon hydride—can be laid down from the gas phase as a thin film, so thin that amorphous silicon cells can be formed on flexible substrates such as stainless steel. These cells are less efficient than crystalline cells, being currently around the eight to 10 per cent mark, but they are also less expensive.

Other semiconductors are occasionally used. Gallium arsenide (GaAs) cells are very efficient and perform well under high illumination; there may be a place in the future space market for GaAs and also for its sister material, InP, although the present market is dominated by silicon. The original Bell silicon cell was developed for use in space, and the first of many satellites to use PV power was Vanguard I, launched in 1958. And in the Science Museum can be seen the flight spare of Prospero, a historic satellite, launched on a Black Arrow rocket from Woomera in 1971 in the only all-British space launch ever. It was a technology satellite, flown mainly to test its solar cells, of which it had a large number. Two years ago, the last time it was monitored, Prospero's PV cells were still powering its transmission system, and they may be talking yet, but no one is listening any more. One a happier note, when the Hubble Space Telescope's faulty optics were recently corrected, its original PV sails were replaced with a double roll-out silicon array, manufactured at British Aerospace in Bristol.

In the 1970s, with the rise in the price of oil, PV modules were first produced specifically for terrestrial applications. Since then, commercial module efficiency has improved to about 18 per cent, and costs have fallen about sixfold. The PV market is growing, with about 60 MWp produced last year, and about 380 MWp cumulatively. PV power is used in a wide range of places and applications. Consumer goods like watches and calculators constitute about one-third of the total market. At the other end of the scale, there are some very large arrays that feed their power to the grid, mostly constructed for demonstration purposes in the early 1980s. PV is often the preferred choice for off-grid applications, such as remote telecommunications stations, navigation lights, and weather stations. PV also brings great benefits to the developing world: electrical power for lights and radio, refrigeration and air conditioning, water pumping, and village electrification.

Closer to home, there are an increasing number of houses, both stand-alone and grid-connected, with PV cells on their Sun-facing façades; in Germany, the government is committed to installing 2000 such roofs. In the past couple of years, PV cells have started to spread to larger buildings in Europe. Such building-integrated PV arrays typically generate some tens of DC kilowatts peak, which is inverted to AC and either used within the building itself or sold to the local grid. The UK will soon get its first PV-clad building, thanks to the tireless efforts of Professor Bob Hill at the University of Northumbria, where a 1960s Computer Studies building will be getting a PV-clad face lift in summer 1994.

Photoelectrochemical solar energy conversion

Now let us turn to a range of direct conversion devices known as photoelectrochemical converters. These are basically electrical batteries that work in the light but not in the dark. They are much less highly developed than PV cells, but they go back further into the nineteenth century, to the French scientist Edmond Becquerel, who in 1839 communicated to the Académie des Sciences his discovery that, if a platinum electrode coated with silver halide was illuminated, a substantial photocurrent was produced. Becquerel experimented with filters of various colours, and concluded that the photocurrent was a measure of the number of chemical active rays of light, throwing in the incautious *obiter dictum* that his discovery rendered obsolete the earlier method of M. Biot of estimating the chemical effect of light of various colours from the tint it produced in a sheet of silver chloride-coated paper.

This drew a magisterial response from the M. Biot in question, one Jean Baptiste Biot, Professor of Physics in the Collège de France, no less, and an authority on the dispersion and polarization of light: '*Tout effet résultant d'une cause physique ne lui est pas pour cela proportionnel.*' ('The magnitude of a physical effect does not necessarily provide a linear measure of its cause.') '*Il est difficile de répondre,*' admitted Becquerel, but respond he did, reiterating his view that the current depended on the number of light rays, not their colour. Biot was back in print the following week. '*Je croirais inutile de répéter mes objections ici.*' Useless or not, he then proceeded to repeat his objections: '*The Academy will readily see that what follows has a purely scientific purpose. I have nothing against a young man whose zeal and inventive mind I appreciate.*' Biot was well away, citing the colour sensitivity of the new Daguerre photographic process in support of his view that the photocurrent was not a question of the number of rays, but of their colour. The controversy rumbled on for years, hardly capable of resolution until the quantum theory of light and electricity provided a natural link between the one photon absorbed by the electrode and the one excited electron it causes to move in the circuit. In a sense, Becquerel and Biot were both right: only those colours that correspond to the semiconductor bandgap energy or above can produce the hole–electron pair and the photocurrent, but above that energy, it is a question of the number of photons, not their colour.

The photocurrent flows because the junction between the illuminated electrode and the electrolyte behaves like a PV junction, bending the energy bands and separating light-generated hole–electron pairs. The Becquerel effect is the basis of a promising photoelectrochemical device, currently being developed by Professor Michael Grätzel at the Ecole Polytechnique Fédérale in Lausanne, using the semiconductor titanium dioxide (Fig. 7). Titanium dioxide is far from exotic—it is the commonest white pigment in paints and paper. It is used in photoelectrochemistry because of its high chemical stability, but the fact that it is white tells us that, un-

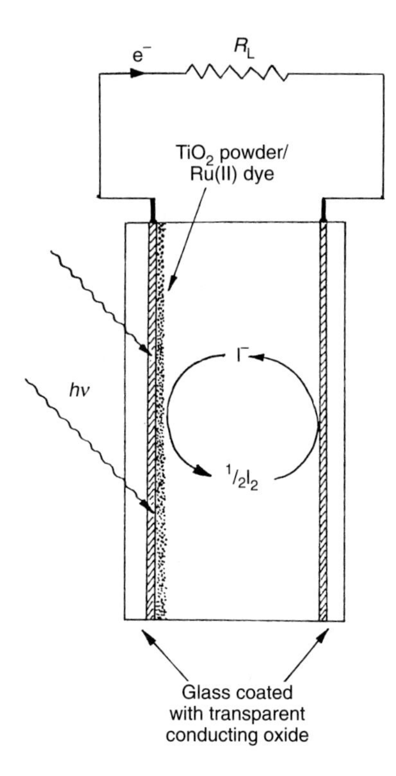

Fig. 7 The Grätzel electrochemical photovoltaic cell. The anode consists of a high surface area titanium dioxide (TiO_2) paste fired onto an optically transparent base and coated with an orange ruthenium dye. The solution contains the iodine/iodide redox couple dissolved in a non-aqueous solvent such as acetonitrile.

fortunately, it does not absorb visible light. So Grätzel made a coloured electrode out of titanium dioxide by firing a paste of it onto conducting glass, and dying the film with a ruthenium derivative. The iodine/iodide couple in solution is oxidized at the dyed electrode, and reduced at the other electrode, so the cell is an electrochemical PV cell, producing electric power rather than a chemical fuel. Grätzel's cells are over 10 per cent efficient in artificial light, better than amorphous silicon and only about one-fifth of the cost, and they are currently under commercial development in Germany.

Solar hydrogen generation

Making electrical power is fine, but what if we could use sunlight to make hydrogen electrochemically? When water is electrolysed in the dark by applying a voltage between two metal electrodes, it is decomposed into its elements, hydrogen and oxygen, and this gaseous mixture stores a good deal of energy. Michael Faraday, who established the laws of electrochemistry at the Royal Institution, liked to demonstrate this by trapping a little of the gas mixture in a soap bubble in his hand. If we put a light to the bubble, we can readily appreciate the potential of hydrogen as a fuel, as did another Frenchman, Jules Verne, over a hundred years ago, in this remarkable passage from his novel *L'Ile Mystérieuse*:

> I believe that water will one day be used as a fuel, because the hydrogen and oxygen which constitute it, used separately or together, will furnish an inexhaustible source of heat and light. I therefore believe that, when coal deposits are exhausted, we will heat ourselves by means of water. Water is the coal of the future![5]

The future moved a step closed in 1972, by the efforts of two Japanese scientists—Akiro Fujishima and his supervisor, Kenichi Honda—working together at the University of Tokyo. They made a crystal of titanium dioxide into an electrode, immersed it in acid and connected it to a platinum counter electrode in alkali (Fig. 8). They then illuminated it with ultraviolet light, and observed that the water was *photoelectrolysed*, that is, oxygen was produced at the titanium dioxide electrode and hydrogen at the platinum electrode, but no external power source was required, only the driving force of the absorbed light.

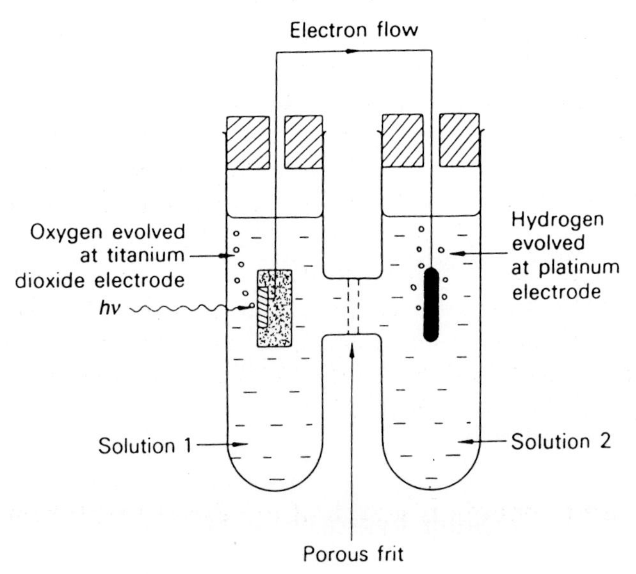

Fig. 8 The Honda–Fujishima cell for the assisted photoelectrolysis of water.

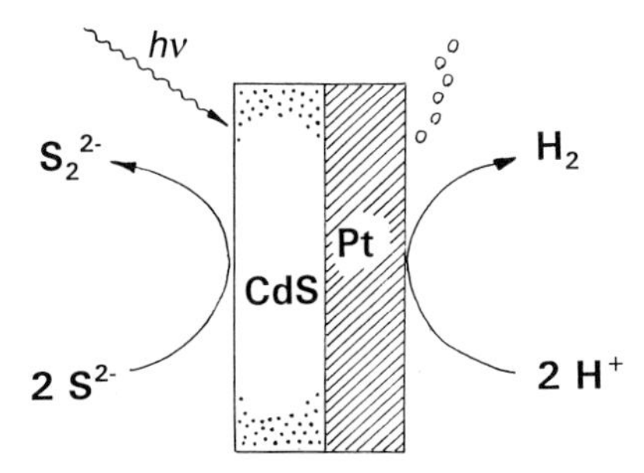

Fig. 9 Nozik's hydrogen-evolving chemical diode.

Unfortunately, there are two problems with the Honda–Fujishima cell: first, as pointed out above, titanium dioxide does not absorb visible light, but only ultraviolet, of which there is little in sunlight, at least while the ozone layer stays in place; and second, the pH difference between the two solutions effectively reduces the energy storage efficiency to well below one per cent, too low to be useful. Fujishima and Honda's announcement therefore started a race to find a semiconductor with better properties for water splitting. There were glorious noonday breakthroughs, shortly followed by false dawns. Alas, semiconductors which absorb visible light efficiently tend to be chemically unstable, which is why Grätzel used a sensitizing dye. So far, a stable semiconductor that splits water efficiently using visible light has not been discovered, and that has led scientists to look at doing each half of the process separately, in the hope of combining them later. Figure 9, for example, shows the hydrogen-producing system devised by Arthur Nozik at the Solar Energy Research Institute in Colorado. It is very simple, consisting of the semiconductor cadmium sulphide, which does absorb visible light, with a platinum counter electrode bonded to its back. There is no need for any external wires for a device that is to be run at short circuit. When this diode is immersed in a solution containing sulphide and illuminated, hydrogen is evolved on the platinum and sulphide is oxidized to sulphur on the cadmium sulphide. Unfortunately, putting such a system together with an oxygen-producing system to make a complete water splitting system in which all other products (such as the sulphur in the Nozik cell) are recycled has not yet been achieved after 20 years of effort.

Solar energy research at the Royal Institution

Twenty years ago, I was lucky to be in the right place at the right time. I had just moved from Oxford to London in order to work with Sir George

Porter (as he then was) at the Royal Institution, switching my field to photoelectrochemistry, in the hope of making myself vaguely relevant to the thrust of his photophysical research. It was here at the Royal Institution that I was instrumental in forming the UK Section of the International Solar Energy Society, and here that the inaugural meeting of the Society was held, 20 years ago.

Chemists are supposed to be good at synthesizing molecules, but some of us are better at destroying them, and indeed, thanks to Superfund, there can be more money in cleaning up pollution than in making a fine chemical. So let us suppose that Solarville's water supply has been contaminated; we can clean it up using the Solaqua process recently developed by Professor Jim Bolton of the Toronto-based company Solarchem. To demonstrate this process a small amount of an environmentally friendly iron complex is added to a portion of the contaminated water, which is then exposed to a light source, for example the Sun. The sunlight generates hydroxyl radicals from the reagent, and these oxidize a wide range of dangerous organic substances such as traces of dry cleaning fluid or dioxins in the polluted water, to harmless carbon dioxide and water. The by-products of the Solaqua reagents are also completely harmless. The treated water gradually turns colourless as it is cleaned up by the Sun.

This example illustrates one of the main driving forces behind the current renaissance of interest in renewable energy—concern about the environment, and in particular concern about the impact of increasing atmospheric levels of carbon dioxide on the climate. There is no doubt that in resource and technology terms, the renewable forms of energy can potentially have a serious impact on this problem, but the question is will they?

In summary, an old Irish proverb seems appropriate: '*Ní hé là na gaoithe lá na scolb*', or '*Don't put your trust in the thatcher on a windy day.*' Naturally I intend no major political pun. The moral is, is it not, that we should get our roofs safely thatched before the wind blows too unkindly. And I hope that I have convinced you that, when the oil wells finally do run dry, the Sun will rise to the challenge.

Notes and References

1. Chatwin, B. (1993) *The morality of things*. Typographeum, 1993, p. 20.
2. This and other quotations were read at the Discourse by the distinguished National Theatre player, Martin Jarvis.
3. Ciamician, G. (1912). *Proceedings of the Eighth International Congress of Applied Chemistry*, Washington and New York, 4–13 September, 1912.

4. US. Patent No. 2,443,542, 27 May 1941, 'Light-Sensitive Electric Device including Silicon', Russell S. Ohl, assignor to Bell Telephone Laboratories, Incorporated.
5. Verne, J. (1874). *L'Ile mystérieuse.*

MARY ARCHER

Born 1944, educated at Cheltenham Ladies' College, St. Anne's College, Oxford and Imperial College, London. Lecturer in Physical Chemistry at Somerville College, Oxford from 1971 to 1972. Research Fellow in the Davy Faraday Laboratory of the Royal Institution from 1972 to 1976 and Fellow and College Lecturer in Chemistry at Newnham College, Cambridge from 1976 to 1986. Has been a Senior Academic Fellow at de Montfort University since 1990, a Visiting Professor in the Department of Biochemistry at Imperial College since 1991 and a Trustee of the Science Museum since 1990. Became a Fellow of the Royal Society of Chemistry in 1987.

Are plants intelligent?

MALCOLM B. WILKINS

Introduction

The year 1993 marked the 250th anniversary of the birth of Sir Joseph Banks, a gentleman of substantial private means who devoted his life to the patronage of science, especially botany. He was, above all else, an explorer and plant collector who, at his own expense, accompanied Cook on his expedition to the southern hemisphere. The magnificent genus *Banksia* bears his name in recognition of his original collection of plants in Australia. Banks was instrumental in persuading the British Government of the economic advantages of acquiring and extending knowledge of the global flora, and of transporting potential crop plants from one continent to another for exploitation. Later in life Banks became probably the greatest 'fixer' the scientific community in this country has ever seen, or is ever likely to see. The King's confidant, and President of the Royal Society of London for more than 40 years, it was Banks who persuaded the Crown to appoint Sir William Hooker to be one of my distinguished predecessors in the Regius Chair of Botany in Glasgow University in 1820. Banks was instrumental in making the Royal Botanic Garden, Kew, the centre for plant collection, cultivation, exchange, and exploitation which resulted in far reaching economic and political benefits to mankind. Banks' vision for Kew was realized in the years following the transfer of Sir William Hooker from Glasgow to be the first Director of the Royal Botanic Gardens in 1840, the year that Banks died. Banks himself published rather few scientific papers, the most significant of these being in the field of plant pathology.

I am delighted that the Royal Institution decided to commemorate Banks' anniversary by a Discourse on plants, because it is an unfortunate fact of life that many people, especially young people, regard plants as being rather dull and uninteresting when compared with animals; 'boring' is a term often used. This situation is perhaps a sad reflection of the failure of plant biologists to project a vibrant image of their subject to the public at large, and especially to school pupils, to whom all too often the study of plant biology is synonymous with the cutting of stem and root sections for microscopical examination. I am always delighted to have an opportunity of dispelling this unfortunate image of botany, and of convincing an audience that plants are amongst the most exciting, important, sophisticated, and interesting of living organisms that inhabit this planet.

There are two quite straightforward reasons why plants are important and intrinsically interesting. Firstly, they alone are responsible for sustaining life on this planet and for providing most of our creature comforts, through their unique capacity to capture and store solar energy. Secondly, they are extremely complex and highly sophisticated living organisms possessing, for example, a full range of sensory systems comparable to our own, and the ability to conduct some of the most complex synthetic organic chemistry known. Finding out how plants work is an exciting activity which poses formidable intellectual and technical challenges.

Our very existence, as well as that of all other animals on this planet, depends absolutely upon the capacity of the green plant to capture and store the Sun's radiant energy, and to carry our a huge number of sophisticated biochemical reactions that provide us not only with our food in the form of carbohydrates, proteins, and fats, but also with our beverages, the oxygen we breathe, and with a wide variety of structural polymers with which we clothe ourselves, communicate with one another, build our houses, and construct all kinds of items to make our lives comfortable. In addition, plants provide a huge range of very complex organic chemicals essential for our health, and with which we alleviate disease and suffering. Our requirement for inorganic ions is met in large measure only through the ability of plant roots to accumulate such ions against a concentration gradient from the soil water. Plants are solar-powered, synthetic organic chemists, and have the added advantage of being the principal contributors to the aesthetic quality of our homes and gardens, and of the natural environment.

The advent of plant genetic engineering has provided plant biologists with the means of modifying the biochemical machinery of plants in highly specific ways and with an unrivalled opportunity of alleviating starvation and malnutrition. The development of crop varieties specifically designed to withstand particular environmental hazards such as drought, disease, high light intensity, low soil nitrogen availability, and low temperature, or to provide a better balance of essential nutrients such as amino acids, is a realistic possibility in the years ahead. Now that gene transfer between plants is almost a routine laboratory exercise, I am in no doubt that Banks, had he been alive today, would have been enthusiastically advocating the potential of these new advances.

What appears to be much less appreciated in the minds of most people is that in order to colonize the harsh terrestrial environment and to capture and store the Sun's energy in an efficient manner, the green plant has had to evolve into a highly complex and sophisticated organism. A wide range of sensory systems has had to be acquired to monitor the surrounding environment, to adjust the growth, development, and orientation of roots, stems, and leaves, to regulate biochemical reactions on a temporal basis, to detect physical contact, and to *anticipate* the onset of adverse climatic conditions. Plants have thus become intrinsically very interesting organisms and it is upon their remarkably sophisticated sensory mechanisms that I wish to dwell in this article.

Among the accomplishments of plants are

(1) the capacity to detect light (effective colour vision), gravity, touch, and chemicals (taste and smell);
(2) the abilities to talk, count, measure time accurately, apparently predict the future, and recognize their friends and enemies;
(3) the possession of a memory and a nervous system.

This is a somewhat dramatic list of the sensory capabilities found in higher plants, and some of the assertions listed will undoubtedly appear to be preposterous, but I shall show by simple experiments that every one of these assertions is valid. I want to convince you that plants have all the sensory capabilities that we have ourselves. If I am successful, it will be impossible for me to avoid the question of whether, like humans, plants make intelligent use of the constant stream of information they are receiving from their sensory organs, or whether their responses to environmental signals are strictly pre-programmed.

Sensory systems in plants

Colour vision

Quite apart from the photoreceptor systems involving the chlorophyll and carotenoid pigments used to capture solar energy for photosynthesis, plants have two other light sensing systems, each of which detects different spectral regions (colours) and regulates specific physiological and biochemical activities in plants. Almost everyone knows that plant shoots bend towards a light source if unevenly illuminated, for example on a window sill. The classical experiments of Darwin in 1880 on the role of light in regulating the direction of growth of the shoots of young grass seedlings revealed the photosensor to be located at the extreme apex. Subsequent studies showed this bending response to be elicited only by light in the ultraviolet and blue regions of the spectrum. Only a few quanta of radiation are required to induce curvature—it is an extremely sensitive system—and the pigment has now been identified as a membrane-bound flavoprotein. Excitation of this pigment system more on one side of a shoot than the other activates a highly specialized and polarized transport system for a growth regulating compound (indole-3-acetic acid or auxin) which accumulates in greater amounts on the shaded side of the shoot, at the expense of the lighted side, thus inducing differential cell growth on the two sides of the organ and hence curvature.

The second photosensing system comprises the chromoprotein pigment phytochrome, which is responsible for the absorption of light in the red (600–700 nm) and near infra-red (far-red, 700–800 nm) spectral bands. This photosystem regulates a great many developmental, biochemical, and molecular processes. There are now known to be at least three phytochromes of different molecular masses; the chromophore is a linear tetrapyrrole molecule and the protein has been sequenced. Phytochromes are unique in that they can exist in two forms, one absorbing red, and the other absorbing far-red, radiation. The two forms can be freely interconverted by exposure to either the red or far-red spectral band. The far-red absorbing form of a pigment is usually the active form for initiating or arresting a developmental process or the synthesis of an enzyme (expression of a gene), but there is now evidence that for one or two processes the red-absorbing form is the active form.

Some responses of plants to exposure to red light are rather slow, requiring several hours to appear. Such responses include the germination of seeds of the light-sensitive, Grand Rapids variety of lettuce, and the change from the etiolated pattern of growth of seedlings in darkness, to the familiar pattern of growth in the light, with expanded leaves and shorter stems (Plate 13). There is no doubt that conversion of phytochrome from one form to another regulates the expression of genes in both a positive and a negative way, thus either inducing or suppressing the synthesis of a large number of enzymes. There is also evidence that phytochrome can induce almost instantaneous effects which are most likely to be attributable to a direct effect on the membranes of the plant cells.

Plants therefore clearly possess two, quite distinct 'eye' systems with which they sense the blue and red ends of the spectrum. These systems provide plants with the essentials of colour vision: the ability to detect and respond to different wavelengths of electro-magnetic radiation. Both plants and fungi also have the capacity to focus light on to photoreceptor pigments.

Gravity

For a seed germinating beneath the soil surface, an absolutely critical requirement is for the newly protruding root and shoot to grow immediately in the correct direction (Plate 14). Unless the root grows vertically downwards the seedling will attain neither secure anchorage nor a reliable supply of water and inorganic ions. Likewise, no seedling will survive unless its foliage leaves reach light before the food reserves of the seed are exhausted, and the quickest route to the soil surface is vertically upwards. Since roots and shoots grow in the correct direction in total darkness on Earth, but not in the effective absence of gravity in orbiting space vehicles, it is clear that they must have gravity detectors. Indeed, this fact was demonstrated in 1704 by Knight who placed seedlings around the edge of a horizontal cartwheel and found that on spinning the wheel the roots grew outwards and the shoots inwards. The detection of gravity, or mass acceleration, is apparently achieved in plants in much the same way as in animals—by the sedimentation of particles. Within certain specialized plant cells there are structures called amyloplasts—groups of starch grains surrounded by a membrane. These structures occur in cells in the root caps and in those surrounding the vascular bundles of shoots. There is considerable evidence that their sedimentation towards the lower side of the cells containing them is critical for the perception of a plant organ's orientation in a gravitational field. Removal of the root cap, or just removal of the amyloplasts from within the cells of the root cap by chemical means, lead to the loss of a root's capacity to detect its orientation in a gravitational field. However, there is a mutant of a small plant called *Arabidopsis thaliana* that appears to lack amyloplasts but can nevertheless respond to gravity. It may be that other subcellular particles can substitute for the amyloplasts in this mutant because its sensitivity to gravity is less than that of the normal plant.

The gravity detection system initiates slightly different rates of cell growth on the upper and lower sides of roots and shoots to give rise to the curvature that establishes the correct direction of growth for the organ, or restores the correct direction if an organ has been displaced. The 'fine-tuning' of the growth rate of the cells on the upper and lower sides of the horizontally placed organ is achieved by the gravity detection system initiating a downward lateral transport of a growth regulating compound towards the lower half. In the case of shoots this regulator is the growth promoter indole-3-acetic acid or auxin, and its asymmetrical distribution in favour of its lower half gives rise to upward curvature of the organ. In roots the chemical nature of the regulator is not known for certain, but it must act as an inhibitor of cell extension on the lower side to achieve downward curvature. The capacity of plant organs to detect deviation from their predetermined direction of growth is very high; a freely growing root deviates by only a few degrees before a corrective bending response occurs.

Sense of touch and nervous systems

Plants are extremely sensitive to mechanical stimulation and respond to being touched in a number of ways. The slightest touch of a pea or passion-flower tendril will, within a few minutes, initiate the complex curling responses that are essential for such climbing plants to cling to supporting structures. It now appears that plant tissues respond to slight mechanical stimulation by releasing calcium ions from internal cellular compartments into the cytoplasm. Some plants are very sensitive indeed and respond instantaneously; for example, the stamens of *Berberis* flowers close together instantly when touched by an inquisitive insect, thus ensuring that the latter is liberally covered with pollen. The stigma of *Mimulus* flowers has two flaps that are normally wide apart, but which close together immediately they are touched, thus ensuring that they receive pollen from a visiting insect. In Australia flowers such as *Stylidium* have extremely violent and instantaneous movement mechanisms to ensure pollination. The aquatic plant *Utricularia*, or bladderwort, catches small animals such as *Daphnia* with a suction trap that is triggered by a single stimulus of a sensory hair.

In a number of species, the effects of a mechanical stimulus are transmitted throughout the plant very rapidly. The best known and most dramatic example of rapid stimulus transmission is in *Mimosa pudica*, but *Biophytum* behaves similarly, only more slowly. If the leaflets at the extreme tip of a *Mimosa* leaf are slightly singed with a match, the leaflets collapse, and then all the other leaflets along the leaf collapse in turn. Finally, the whole leaf stalk will collapse as the stimulus passes to a specialized motor structure (pulvinus) at its base. The stimulus may then be transmitted up and down the stem to other leaves. The rate of transmission of the stimulus along the leaf can be determined by timing the collapse of successive leaflets and is much too fast to be attributed to the diffusion of a chemical. In fact, stimulus transmission can be shown to involve an electrical mechanism not dissimilar, in principle, to that which occurs in human nerves. Sensitive electrodes placed in the central stalk of the leaf reveal the cells to have a 'resting' potential of about −150 mV across their outer membrane. When the stimulus passes, this voltage instantaneously disappears but is gradually re-established after about 2 min. This 'action potential' is of almost exactly the same shape as that found in mammalian nerves when a stimulus passes along the cells, though in nerves the resting potential is re-established very much more quickly. In nerve cells the change in potential is due primarily to the movement of sodium ions across the cell membrane, whereas in plant cells, potassium and calcium ions appear to be involved.

Counting and memory

Some very specialized and sophisticated touch sensors are found in the Venus fly trap, *Dionaea muscipula*, a plant that catches and digests animals to acquire its supply of nitrogen since it lives in habitats in which the soil is very deficient in this element (Plate 15). These sensors take the form of six multicellular hairs, three on each face of the trap. The extraordinary thing about this plant is that it can count and has a memory. This is clear from the fact that if a sensory hair is touched once, the trap does not close, whereas a second touch applied to either the same hair, or to any of the other hairs, causes the trap to close instantly. This plant can thus distinguish between not being stimulated, being stimulated once, and being stimulated twice—a clear capacity to count up to two. It must also possess a memory, because it is only possible to respond to the second stimulus and not the first one if it can remember that it has already received the first stimulus. Experiments in which the interval between the two stimuli is varied show that the memory lasts for about 1 min. The fact that any of the sensory hairs can be stimulated, and that the responding motor cells in the middle line of the trap are about 10 mm distant from the hairs, means that there must be an extremely rapid communication system between the hairs and the motor cells. An electrical transmission system is again involved: propagation of 'action potentials' can be recorded from the base of the sensory hair to the motor cells through the cells on the surface of the trap. The sensory hair is, in fact, a mechano-electro transducer, converting a mechanical stimulus into an electrical one. This conversion is achieved by the stiff upper part of the hair acting as a lever which, when displaced, squashes large pressure-sensitive cells at the base. These cells immediately respond by firing an 'action potention', which is then transmitted to the motor cells. The location of the memory mechanism is not known, although it seems most likely to be in the responding motor cells. Precisely how the motor cells move the trap so quickly is also a matter of some doubt.

Taste

In common with a number of carnivorous plants, the Venus fly trap has chemo-sensors, or taste buds, on its surface to detect the chemical nature of its prey. If these sensory glands detect nitrogen-containing amino acids or proteins, they initiate two processes: a further slow closure of the trap to press the two halves against the prey, and a secretion of proteolytic enzymes that digest the prey to release soluble, nitrogen-containing compounds which are then absorbed into the plant. If the taste buds do not detect such molecules then the trap opens again after about 25–36 h; the trap does not try to digest dead leaves and other material containing largely carbohydrate polymers.

Talking

The proposition that plants can talk sounds rather far-fetched but it is nevertheless true that plants do emit a specific sound under particular circumstances. Sounds emitted by human beings mean absolutely nothing if the language is not understood, and similarly the sound emitted by a plant can only be interpreted if the cause of its generation is known.

In the normal course of events, water enters plant roots from the soil, and is drawn up the roots and the stem in specially adapted, thick-walled tubes comprising the xylem tissue, the wood of tree trunks, eventually to be lost from the leaves in the process of transpiration. If the soil in which a plant is growing becomes dry, and water continues to be lost even at a very low rate from the leaves, the plant becomes water stressed and begins to emit a clicking noise which can be detected with a sensitive microphone and appropriate amplification equipment; it cannot be heard with the unaided ear. The noise is generated by the continuous micro-columns of water that run from the root tips to the shoot tips in the xylem cells being stressed or stretched to the point where they break or cavitate. Each cavitation event gives a click. Thus the clicking noise in effect is the plant indicating a need for water.

The potential for a plant to control its own watering schedule in a glasshouse through its capacity to indicate audibly when it is water stressed has already been investigated. When water is supplied the plants become silent so it does not require very sophisticated electronic equipment to provide water to clicking plants and not to silent ones. Thus, plants can ask for water—if you understand how the sound arises or understand the 'language'. Whether plants emit other sounds in response to other stimuli or situations, is not yet known.

Plate 13 The pattern of growth of French bean seedlings in darkness (left) and light (right). Note the long stem with a hooked apex and unexpanded leaves in darkness, and the straight, shorter stem with expanded leaves in light.

Plate 14 A dissected germinating seed of sweetcorn, *Zea Mays*, showing the embryo with clearly defined shoot and root which would soon have emerged from the seed. The storage tissue (endosperm) is stained black, indicating the presence of starch. The endosperm is the energy source for the young seedling, but in this and other cereals comprises our most important food.

Plate 15 A Venus fly trap showing the six red sensory hairs, three on each face of the trap.

Time measurement

That living organisms have the capacity to measure time was discovered in plants more than two centuries ago from the persistence of rhythmic movements of leaves for many days after plants had been isolated from environmental variations in a darkened basement. Biological clocks are now known to exist in virtually all living organisms, including prokaryotes, and underlie such annoying phenomena as jet lag in ourselves.

The phenomenon of photoperiodism was discovered in 1920 in studies of the control of flowering in plants. Many plants develop flowers only at specific times of year and the accuracy of this timing is quite remarkable. Some, such as *Chrysanthemum* and *Kalanchoë blossfeldiana*, flower only in the short days of winter, whereas others, such as *Hibiscus* and *Hyoscyamus niger*, flower only in the long days of summer. The description of such plants as short-day and long-day plants proved to be a misnomer because the response has nothing whatever to do with day length; rather it is the length of the night which is critical. Plant photoperiodic responses are very complex in that they involve the measurement of the length of time which elapses between two light signals at dusk and dawn, the photoreceptor involved in the perception of these signals being none other than phytochrome, mentioned earlier. Whether a plant flowers is determined by it detecting whether or not the length of uninterrupted darkness exceeds (in the case of short-day plants), or is less than (in the case of long-day plants), a critical value, the critical night length, which is different but specific for each species and can range from a few to 16 or more hours.

In order to measure night length accurately it is necessary to have a 'clock' mechanism, that is an oscillator which does not vary in its frequency of oscillation with changes in environmental factors, particularly temperature. This was a major problem with early mechanical clocks where the pendulum rod was made of iron. Such clocks gained time on cold days and lost time on warm days due to the thermal expansion of the iron. Elaborate compensated pendula were devised to ensure that there was no change in length with temperature. In plants, oscillating systems exist which astonishingly have this clock-like feature–the period of oscillation is temperature compensated to a very high degree—so that they can form the basis of an accurate time measuring system. These oscillations are manifest in a large number of behavioural, physiological, and biochemical phenomena in plants, animals and micro-organisms, with the circadian (from *circa*, about; and *dies*, a day) rhythms in leaf movement, the rate of synthesis of specific protein components of the chloroplasts, the opening and closing of stomata, the activity of certain key enzymes involved in the assimilation of carbon dioxide, and the synthesis of enzymes (expression of specific genes) being particularly well-studied examples in plants. These rhythms indicate the very precise temporal control the basic clock system has over the biochemical reactions in plant cells, a control which in many cases is essential to prevent futile cycling of metabolic systems and a consequent waste of energy.

Predicting the future

One consequence of plants having the capacity to measure time is the apparent ability to predict the future! Everyone is familiar with the winter buds which are so prominent on our trees in January and February. These are specialized protective structures to ensure that the delicate stem apex and its attached young foliage leaves for the next season are protected from damage by the low temperatures of winter. The development of the winter bud involves a total change in the pattern of leaf development involving the cessation of foliage leaf production and the onset of scale leaf production. The latter are structurally, functionally, and biochemically totally different from foliage leaves and involve the expression of a quite different set of genetic information. However, if trees are examined in July and August, it will be noticed that the winter buds are already fully formed, clearly indicating that the plant is *anticipating* the advent of adverse climatic conditions some 4–5 months in the future. How is this predictive capacity achieved and the appropriate protective structures developed? The answer again appears to be the capacity of the plant to measure night length—short nights of summer initiate winter bud development so that by the year's end the plant is fully prepared to meet the rigours of January and February.

Recognition of friends and enemies

Finally, plants have the capacity to recognize their friends and enemies. Survival for both plants and animals depends utterly upon the capacity to recognize enemies, especially pathogens, so that appropriate defensive mechanisms can be activated. There are a great many situations in which plant recognition systems are of significance, and in virtually every case complex chemical signals are involved. In many cases these signal chemicals are glycoproteins, that is proteins with various sugar molecules attached, in this respect plant cells use surface recognition signals similar to those of animals.

Plants apparently use recognition systems in a wide variety of situations. Each gamete (sexual reproductive cell) of the alga *Chlamydomonas* bears recognition systems on the ends of its two flagella. The signals indicate not only whether the gamete is of the plus or minus form (not male and female, because the gametes are morphologically identical) but also the species of *Chlamydomonas* to which the gametes belong. Fusion of gametes only takes place between those that recognize each other as being plus and minus of the same species.

There are many situations where different plants and micro-organisms set up some kind of relationship with each other, often to mutual benefit, and the specificity of the species or strains of the partners involved implies that recognition mechanisms are involved. Lichens comprise specific species of algae and fungi living together to mutual benefit; the establishment of nitrogen-fixing root nodules in leguminous plants requires a particular strain of the bacterium *Rhizobium* with a particular species of legume; the parasitic habit of such flowering plants as *Cuscuta* and *Orobanche* is only established with certain other flowering plant species; and, as horticulturalists know well, the grafting of plants is only possible between closely related species. All these phenomena imply the existence of signals on the cell surface, so that when cell to cell contact is established, recognition takes place and a relationship is either established or not.

One of the best studied and important cases of recognition signals in plant cells concerns those on the surface of pollen grains. These take the form of glycoproteins that may be derived from both the mother plant tissue (tapetum) or the pollen grain cells themselves. They are capable of initiating immunological reactions in humans, being the cause of hay fever.

The signals on the surface of pollen grains play an important role in preventing self-fertilization in many species of flowering plants, most of which are, in contrast to animals, hermaphrodite. Such plants thus recognize their own pollen when it falls on the stigma and, as a result, biochemical reactions are initiated to produce a special sugar polymer known as callose which effectively smothers the pollen grain and its protruding pollen tube to prevent its growth and the possibility of it fertilizing the egg cells in the ovary. There is evidence that recognition systems are also involved in preventing many intra-specific and intra-genetic crosses, although in the latter case a variety of factors may be involved in preventing the pollen grain from effecting fertilization.

Recognition systems and other mechanisms that prevent the interbreeding of different species and genera of plants can be a serious problem to breeders anxious to improve crop plants. The inability to introduce into a particular crop species a desirable feature from another unrelated crop plant, or even a wild or weed species, has highlighted the need to overcome the barriers which exist when using natural pollination procedures.

In recent years these barriers have been overcome in many instances by using the techniques of plant molecular biology. Their use represents the latest attempts to attain the objectives for which plant breeders have been striving for centuries. What these modern molecular techniques allow is the transfer of specific genes or groups of genes from one plant to another rather than the wholesale mixing of genes inherent in the traditional breeding procedures. In other words, the era of designer plants has begun.

Designer Plants

The most promising and successful method of transferring genes from one plant to another has been to utilize the soil bacterium *Agrobacterium tumefaciens* which normally causes cancerous tumours in plants. The bacterium induces the tumours by infecting the cells of the plant and incorporating some rather unpleasant genes from one of its plasmids into the chromosomes of the plant cell. The affected plant cells thus become permanently altered—transformed—and continue to divide and grow in an uncontrolled way, producing huge tumours characteristic of crown gall disease even when the bacterium is no longer present.

Before it can be used in plant biotechnology, the bacterial plasmid is first disarmed by removing the tumour-inducing genes. The plasmid can then have inserted into it the desirable gene, perhaps one that conveys disease resistance, drought tolerance, or resistance to insect attack. When a bacterium containing this modified plasmid attacks the cultured cells of a crop plant, the new gene is inserted permanently into their chromosomes. As these newly modified, or transformed, cells divide and grow, every new daughter cell will contain the newly inserted gene.

In many plant species, it is possible to regenerate whole plants from such transformed single cells, and so every cell including the reproductive cells in such a regenerated plant will contain the new gene, and possess the desirable new feature if the gene is adequately expressed. The most spectacular demonstration of the success of this procedure has been the transfer of the gene for the enzyme luciferase from fireflies into tobacco plants. Luciferase is the enzyme that causes a substance called luciferin to break down with emission of bright flashes of light in the last three abdominal segments of fireflies. The presence and expression of the luciferase gene in each of the cells of the transformed tobacco plant is amply demonstrated by soaking the plant with a solution of luciferin in the dark when it begins to emit light. While no one can really claim to want such 'self-lighting' tobacco, transfer of this particular gene into a plant dramatically demonstrates that the procedure is successful and can be exploited for the benefit of mankind. Plants have already been engineered to resist insect attack, and to enhance the shelf-life of tomatoes by delaying over-ripening and mushiness which are due to the breakdown of cell wall components.

However, advancement in this field is not without its problems because not all plants are attacked by *A. tumefaciens*, and in some groups of plants successful culture of undifferentiated cells and the regeneration of whole plants from such cells is either very difficult or has not yet been achieved. The grasses and cereals are particularly difficult in both these respects, although progress has been reported recently with regeneration in rice, maize, wheat, and barley. Since the cereals are the most important crop plants for providing us with nutrients, the transformation of these species is a matter of very great importance. For example, improvement of the amino acid balance in maize, particularly in respect of the three amino acids in which it is deficient, and the lack of which give rise to kwashiorkor in children who live virtually entirely on corn meal in some parts of the world, would surely be a priority in anyone's list of transformations to be attempted.

While techniques such as somatic cell hybridization and the use of viruses as vectors for gene transfer in place of *Agrobacterium* have both met with some success, another most promising procedure involves quite simply shooting genes into either the growing tip of a plant or into young embryonic plants dissected from normal seeds. The genes that are to be transferred are prepared and coated on to the surface of minute spheres of gold or titanium 4 μm in diameter. This 'shot' is then loaded into a 0.22 inch calibre cartridge on a plastic disk and fired at the plant tissue with a modified 0.22 inch rifle. The shot penetrates the cell walls and enters the cells, and the genes are incorporated into the plant's chromosomes. This procedure has been successful in transforming cereals and also woody plants which are also at present difficult, if not impossible, to tissue culture and regenerate. Shooting plants (known in the technical journals as transformation by particle acceleration) is thus a very real method for genetically engineering plants that are either resistant to infection by *Agrobacterium* or cannot be cultured and regenerated. The cells transformed at the plant apex will continue to grow and develop into a transformed shoot tip that will in time produce pollen and egg cells from which further generations of transformed plants may be generated.

Conclusion

The range and sophistication of the sensory systems which I have briefly described in this article correspond closely to our own. It is not unreasonable, therefore, to enquire whether plants show evidence of any intelligent use of the information their sensory systems provide. At the present time there is no evidence whatsoever that plants exercise any choice, or any intelligence, in responding to external signals. Their responses are totally pre-programmed. The answer to the question posed by the title of this article is therefore 'no', and some of you may feel a little disappointed having secretly hoped that the answer would be 'yes' or even 'maybe'. The title was of course designed to enable me to tell you about the sensory physiology of plants. I hope that, by approaching plant sensory physiology in a rather novel and unusual way I have convinced you that plants are very much more complex, sophisticated, and exciting than you had perhaps realized hitherto.

Further reading

Galston, A. W. (1994). *Life processes of plants.* Scientific American Library. W. H. Freeman, New York and Oxford.

Wilkins, M. B. (1988). *Plantwatching.* Facts on File, New York and London.

MALCOLM B. WILKINS

Born 1933, educated at Monkton House School, Cardiff, and King's College, University of London. Lecturer in Botany, King's College, London (1958–64), Rockefeller Foundation Fellow, Yale University (1961–62) and Corporation Research Fellow at Harvard University (1962–63). Formerly, Professor of Biology, University of East Anglia (1964–67) and Professor of Plant Physiology, University of Nottingham (1967–70). Appointed by H. M. The Queen to the Regius Chair of Botany in Glasgow University, 1970, becoming Dean of Science (1985–88). Has served on the Governing Bodies of several Government Research Institutes including the Hill Farming Research Organisation in Edinburgh and the Glasshouse Crops Research Institute in Littlehampton, and on Committees of the SERC and British National Space Centre. Is Chairman of the Board of Trustees, Royal Botanic Garden, Edinburgh, a former Chairman of Life Sciences Advisory Committee, European Space Agency, a Member of its Microgravity Advisory Committee and the Lifesat Science Advisory Committee of NASA. Honorary Member, American Society of Plant Physiologists, Fellow and Vice-President of the Royal Society of Edinburgh, member of the Court of Glasgow University. Darwin Lecturer, B.A., 1967. His principal research interests are the mechanisms of time-measurement in plants and in the sensory systems with which they detect environmental stimuli. Author of *Plantwatching* and numerous scientific papers; Editor of two plant physiology textbooks, and a Managing Editor of *Planta*. Has made many radio and television broadcasts to popularize experimental plant biology.

Artificial hearts: from technology to science

IWAO FUJIMASA, MD, D.M.Sc.

Introduction

The artificial heart is familiar as a sophisticated piece of medical technology which can save a patient from desperate heart failure. We already know that humans can survive for 622 days with an artificial heart.[(1)] Today, heart surgeons have applied artificial heart systems to temporary use for post-cardiotomy cardiogenic shock and as bridges to heart transplant. More than 1700 patients had temporarily been given some kind of artificial hearts before 1992. More than three hundred patients have had heart transplants and more than 60 per cent of these have been discharged successfully (Tables 1 and 2).

Akutsu and Kolff started their research and development of artificial hearts in Cleveland Clinic in 1957[(2)] and Atsumi and Nose started their research in Japan in the following year. So thirty-five years have already been spent on this research. The process of developing an artificial heart

Table 1. Combined registry of artificial transplant for post-cardiotomy cardiogenic shock (American Society of Artificial Internal Organs Registry, December 1992)

Type of artificial heart	Artificial hearts transplanted	Weaned (%[a])	Discharged (%[b])
Left ventricular assist	587	299 (50.9)	161 (27.4)
Right ventricular assist	160	63 (39.4)	39 (24.4)
Bi-ventricular assist	450	184 (40.9)	100 (22.2)
Hybrid bi-ventricular assist	26	6 (23.1)	2 (7.7)
Total	1223	552 (45.1)	302 (24.7)

[a] Per cent of weaned patients = number weaned/number of artificial heart transplant patients.
[b] Per cent of discharged patients = number of discharged/number of artificial heart transplant patients.

Table 2. Combined registry of artificial heart (AH) transplant for bridge to heart (NH) transplant (American Society of Artificial Internal Organs Registry; December 1992)

Type of artificial heart	Artificial hearts transplanted	NH transplanted (%[a])	Discharged (%[b])
Left ventricular assist	166	119 (71.7)	107 (89.9)
Right ventricular assist	5	2 (40.0)	0 (0.0)
Bi-ventricular assist	155	104 (67.1)	70 (67.3)
Hybrid bi-ventricular assist	34	12 (35.3)	6 (50.0)
Total artificial heart	191	135 (70.7)	66 (48.9)
Total	551	372 (67.6)	249 (66.9)

[a] Per cent of NH transplanted patients = NH transplanted patients/AH transplant patients × 100%
[b] Per cent of discharged patients = discharged NH transplant patients/AH transplant patients × 100%

is similar to that of developers other life support systems. After some success, the technology applied to patients is in the very early stages of the development,[3] but even now the number of clinical applications is still below 100 per year. This is because artificial hearts cannot stop beating even for only a minute, and moreover should continue to beat for at least five years. This means that the artificial heart pump must be able to beat more than 200 million times. Many clinicians have hesitated about using artificial hearts permanently because they doubt their durability and fear that the patient's quality of life will be low after transplanting an artificial heart.

Development of the tether-free, totally implantable artificial heart is not complete yet. We have designed and manufactured artificial hearts with completely different structure and energy conversion principles from our own hearts. Biomedical engineers have developed the artificial heart from the intention that the artificial heart would have the same physiological functions as the natural heart. On the other hand, physiologists have not paid attention to the technology because such technology is far beyond their imagination. There still exists a great gap between technology and science in this field. In the article, I describe the new technology of artificial heart systems and also some new lines of scientific research initiated by the technological developments.

The heart as a pump

Most people have some idea of the heart's structure and function, but few people except clinicians and basic circulatory physiologists know the precise function of the heart. In order to develop an artificial heart, we must first acquire fundamental knowledge about the mechanical and hydraulic functions of the heart.

The heart is an essential organ that supplies oxygen and nutrients to all organs and tissues and also conveys carbon dioxide and other waste products from tissues. It acts as a pump, sending a fixed volume of blood to all tissues. Two fundamental parameters of the heart are cardiac output (the fixed volume of blood supplies) and ventricular pressure (the constant pressure head needed to push the blood volume against vascular resistance). For vertebrates, the cardiac output increases proportionally to the body weight, but ventricular pressure is almost constant regardless of body weight. As the cardiac output is usually almost 0.1 ml/kg body weight/min, a person who weighs 70 kg has a cardiac output of 7 litres/min. Peak blood pressure is 110 mmHg and mean pressure 80 mmHg. When we make a pump for systemic circulation, then we must prepare a hydraulic pump which can push 420 litres of blood volumes per hour against a water head of 1.5 m. The pump has almost the same function as the water pump of a household washing machine.

When heart substitutes were first being developed, scientists and engineers thought that the key to a successful life support system would be to supply enough blood flow through the capillary network. If enough oxygen-saturated blood could be supplied to the capillary network, humans might survive for an hour or more. This concept guided the invention of a heart–lung machine.[4] A simple 'blood squeeze' tube pump, usually called a 'roller pump', and an oxygenator, can maintain human life for a few hours without heart beating or breathing. Such pumps and oxygenators already existed in the early stages of artificial heart development and influenced the design of artificial hearts. This machine was a very simple prosthesis of cardio-pulmonal function, but it could not maintain human life for more than a few hours. Many physiologists and surgeons thought that this was because the blood flow produced by a roller pump was non-pulsatile. Thus, from the beginning artificial hearts were designed as pulsatile flow pumps.

Our heart consumes only 1 W of energy at rest. A man-made 1 W electric motor is very tiny. But to make a pulsatile hydraulic pump which has pump output of 7 litres against 100 mmHg pressure, we must prepare an electric motor of more than 30 W power input. During exercise, cardiac output is more than five times greater than it is under rest conditions, but the input energy remains less than 10 W. A man-made pump cannot achieve such a high efficiency:size ratio as the natural heart does.

History of the artificial heart

In the early era of artificial heart development, which started in 1957, many types of artificial heart pumps and driving mechanisms were developed. By 1965, more than 30 kind of artificial heart had been designed in the USA, Japan, and Europe. Today, only three types of artificial heart systems remain which we can use clinically: pneumatic, electric, and magnetically driven. Recently, though, new technological lines have been developed and tested in experimental animals. These are completely new concepts, such as an artificial heart made of skeletal muscles, revivals of non-pulsatile flow pump applications, and many small mechanical pumps.

Pneumatically driven hearts were used in the first human applications. This was because in these systems the blood pump and the driving unit were completely separate and so the small blood handling parts were easy to design and manufacture. These systems were easier to control than others because enough power could be supplied by a large pneumatic power source set outside of the body. Today, experimental animals with their hearts completely replaced by artificial pneumatically driven pumps can survive for nearly one year (Fig 1).[5,6] Based upon these experimental results, NHLBI of USA permitted the implantation of a pneumatically driven heart (Jervik 7) permanently in to five patients who were unsuitable for transplants. The Ministry of Health and Welfare of Japan permitted the clinical application of this system as a 'ventricular assist device' (VAD; Fig. 2). The temporary application of a total artificial heart system is usually called bridge use; the number of patients who were awaiting heart transplants with bridge artificial heart had risen to more than 189 cases by 1991. Seventy per cent of bridge use patients can have heart transplants after an average of 24 waiting days (the longest waiting period was 438 days) and 35 per cent of the patients implanted with artificial hearts still survive.[7] However, bleeding, infection, renal failure, and neuropathy have been reported as major side-effects of the total artificial heart implantation[8] and so less invasive VADs have become applied more frequently for bridge use.

The pneumatically actuated pump has a relatively large driving machine outside of the body. If a tether-free and completely implantable artificial heart is to be developed, much more compact driving system must be used. An electric motor is a potential mechanism of directly driving an artificial heart. A brushless DC motor-driven artificial heart has been developed at the *Hershey* Medical Center and a calf with this heart survived for more than one year recently.[9] In this system, a ventricle is

Fig. 1 A goat that survived for 360 days under automatic feedback control. The control objective function for feedback was modified from the value of total peripheral resistance of the goat and the driving parameters were regulated almost every beat. At feeding, urinating, defecating, standing up, sitting down, and running, the goat's blood pump output increased or decreased and its responses seemed physiologic.

attached to each end of a ball and screw shaft which penetrates a DC motor and slides from left to right; the two ventricles are driven reciprocally depending on whether the rotation of the motor is clockwise and counterclockwise. Many other types of totally implantable machines have been developed and some of them were nearly ready to be used in clinical trials, but today the Hershey pump is the only one remaining of them and it is still at the animal experiment stage.

As an alternative to the totally implantable artificial heart, (VADs) have been frequently used clinically. Cardiac surgeons use many pneumatic and also mechanically driven hearts to assist failed cardiac function. Left-side VADs (LVADs) have become particularly popular. Almost 80 per cent of patients with LVADs can go back home successfully after bridge LVAD transplant. Totally implantable LVADs have frequently been used to improve patients' quality of life in bridge use. A DC motor-driven pusher-plate heart (*HeartMate®: Thermo Cardiosystems*[(10)]) and a solenoid-driven pusher-plate heart (*Novacor Medical*[(11)], have been developed for clinical use and applied to bridge patients.

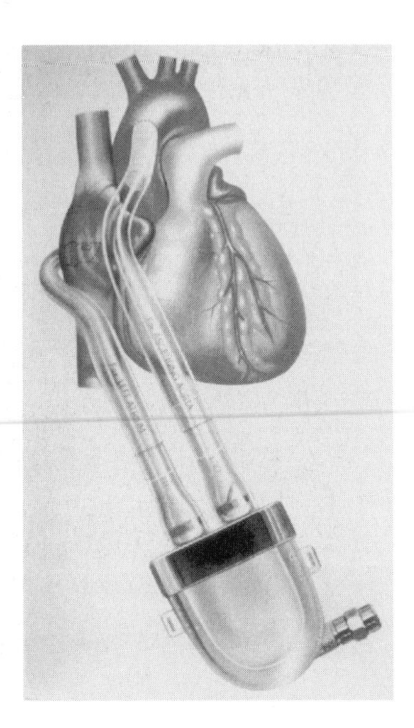

Fig. 2 The first commercially available artificial heart system. Three pneymatically driven sack type blood pumps made from polyvinyl chloride coated with Cardiothane® (ventricular volume 60 ml; fabricated by Nihon Zeon Co.) and a pneumatic driving unit (Corart® 103, manufactured by Aishin Co.) Scheme shows left ventricular cannulation of a ventricular assist pump (LVAD); the inflow cannulation is located in the left atrium or apex of left ventricle; the outflow cannulation is located in the ascending aorta, descending aorta, or abdominal aorta.

In the history of artificial heart development, various approaches have been taken and some unusual research carried out. One approach a scientific challenge to human physiology which started from some clinical evidence. It was the application of continuous flow pumping to permanent artificial heart experiments. The reasons why vertebrates have a pulsatile pump as a heart have never been discussed scientifically. But since a calf has survived for a month with a continuous flow heart pump,[12] we doubt that pulsatile flow is essential for vertebrate life. There may be only two factors that are essential for artificial heart design: cardiac output and pump outflow pressure. Another system that has been developed is a hybrid artificial heart. In the early stages, a pump cylinder was designed that was driven by patient's leg muscle which stimulated electric pulses. Recently, patient's skeletal muscle was applied to driving an artificial heart and to LVADs.[13,14]

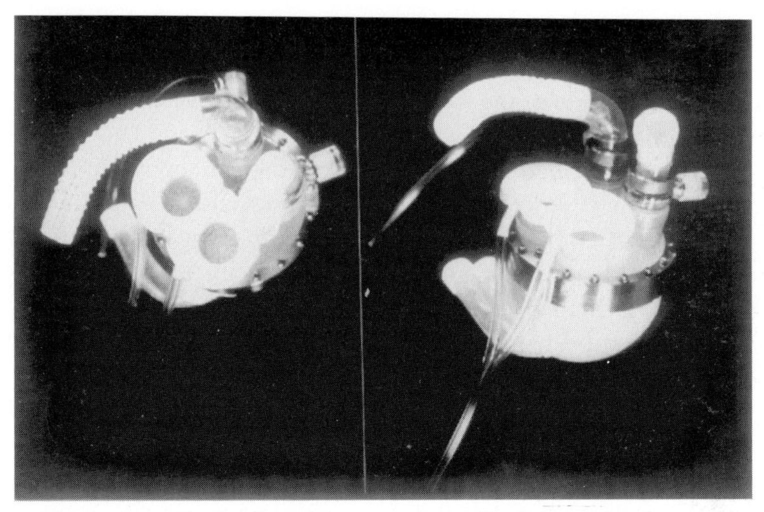

Fig. 3 A totally implantable, pneymatically driven blood pump made from polyvinyl chloride coated with Cardiothane®: atrial cuffs are made from EPTFE covered with Cardiothane®. Four 'jellyfish' prosthetic valves are used. An electromagnetic blood flow meter is inserted between the left ventricle and the left outflow cannula for controlling pump output. (Designed and fabricated by K. Imachi.)

Technology to science

Artificial heart research started out with a very simple reasoning, namely that the heart is a pump, and that by imitating this function, it would be possible to make an artificial heart. Since 1957, artificial heart researchers have sought many possible applications of technology from the engineering field. Three big bottle-necks became clear during the development, and these produced three new scientific fields for us: finding a physiological basis for artificial heart control, developing anti-thrombogenic materials which include clearing hydrodynamical causes of thrombus formation, and designing a novel energy conversion system. These fields did not exist previously in science or in conventional technology. By beginning technological research, we approached many scientific problems.

Physiology of the artificial heart

Many physiologically unknown facts appeared during the first stage of artificial heart experiments. The first question was how to determine artificial heart output properly. When implanting a total artificial heart, all neural connections with the central and autonomic nervous systems are cut off, so there are no feedback loops except humoral and hydrodynamical communications between an artificial heart and the body. How, then, can an artificial heart be regulated?

At first, many researchers imitated the hydrodynamical data obtained in experiments on isolated hearts. Starling's law, which states that cardiac output is proportional to the blood volume entering the ventricle, was applied as a golden rule. If an artificial ventricle pushed out all of its venous return (which is equal to the returning blood volume into the ventricle) in each beat, the pump output might be automatically maintained at physiologically normal values. This seems to be true in isolated heart experiments. Until 1975, many researchers had applied the Starling's law.(15) It was easy to decide driving conditions automatically. However, the cardiac output of the natural heart is strongly controlled by the homeostasis of the body. When a relatively large pump is used in an animal experiment, its output could easily become 1.5 or 2 times higher than the physiologically normal range. As a result severe circulatory collapse usually happened within 3–7 days. This so-called high output syndrome frequently occurred when a large ventricular volume pump was installed. When pump output was limited to 80–100 ml/kg/min in rest conditions, in order to prevent the phenomenon, high output syndrome was never experienced.(16) Even in the first human case, though, this phenomenon arose. The patient's cardiac output had remained steady at 3–4 litres/min for many years, but after a total artificial heart was implanted, the pump output suddenly increased up to 10 litres/min and the patient became severely shocked.(17) The same results, showing that the systemic circulation was within physiologically normal limits, were obtained from an experiment in which left heart output volume was controlled to maintain normal aortic pressure.(18) The experimental animals and the patients lived almost for one year using this fixed control mode.

But this control method is not truly physiological. We have experienced many patho-physiological side effects in clinical cases and animal experiments, which are usually called 'artificial heart syndromes'. Elevation of central venous pressure, high blood pressure, anaemia, and low kidney function are frequently reported side-effects of animal experiments and also chronic clinical cases. In order to prevent such phenomena and to respond to physiological demand, some physiologists have tried to obtain physiologically normal feedback control methods. They have tried to use and validate many kinds of physiological parameters as feedback parameters, includes lactic acid concentration in blood, arterio-venous oxygen gradient, and aortic blood pressure. If we also wish to use exercising status as a controlling factor, the control protocol becomes more complicated and difficult. Some researchers have added body movement data measured by an accelerometer while others have used present high pump output driving conditions for exercise. But with these control methods, the output of artificial hearts never responds as the natural heart does. Moreover, in normal animals, cardiac output fluctuates according to many factors, such as changing body position, eating, defecation, alarm, and especially exercising. Some researchers think that it could be possible to determine ventricular stroke volume and beating interval from a decision table or a formula calculated from some circulatory dynamic parameters obtained from measurements taken on the previous heartbeat. In order to achieve such a system, we have tried to change the driving condition from beat to beat with several objective functions using a completely dynamically responding computer and an artificial heart system with beat-to-beat controllable features. After many trials, we have finally approached a sophisticated, fully automatic control method (Figs. 4–7). The decision functions include admittance of peripheral circulation as a main parameter. The animal can survive for almost a year with the control method and does not suffer from any artificial heart syndrome.

Materials and design

One of the most important technological challenges is to find and to develop antithrombogenic materials. This is the main problem confronting developers of an artificial heart pump. In the early days, the heart pump was made of materials that living organisms recognize as foreign substances. The blood coagulation system senses that the inner surface of a blood pump is foreign and so the coagulation process is activated. In early stages of artificial heart development, severe thrombus formation occurred on the inner surfaces of the pumps because the lack of compatibility of the material with the blood, and the implanted animals did not survive for more than 100 hours.(19) Completely anti-thrombogenic material has still not been developed today. But this problem gave rise to a new scientific field: on the blood–material interface.

In the mid-1970s, many polymer scientists became involved in artificial heart research. After better anti-thrombogenic materials had been developed, the length of survival of artificial heart replaced animals suddenly improved. Biomer® (a type of segmented polyurethane(20), Avcothane® (a copolymer of silicone and polyurethane(21)), and 'biolized' materials (a lining of endothelial cells on a rough polymer surface) have been the most important materials for achieving low thrombus formation in the blood pump. Before that, medical grade silicone rubber was the best anti-thrombogenic material. Many blood-compatible materials have since

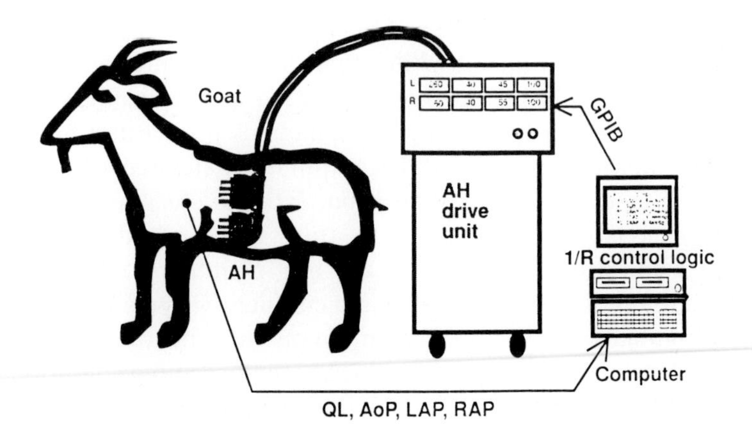

Fig. 4 Feedback loop between an artificial heart goat and an artificial heart drive unit. Left pump outflow volume, aortic pressure, left atrial pressure, and right atrial pressure are measured dynamically, and are fed to a microcomputer. Driving parameters are calculated using an objective function which is modified by total peripheral resistance and artificial heart driving parameters change rapidly. The function is

$$CO(n+1) = CO\,[AoPset(n) - RAPset]/(AoP - RAP) + CP\,[AoP - AoPset(n)]$$

$$AoPset(n) = (1-p)\,AoP + p\,AoPset(n-1) \quad p = \exp(-t/\tau)$$

Where $CO(n+1)$ is the next pump output, CO the measured pump output, $AoPset(n)$ the setpoint of aortic pressure, RAPset the setpoint of right atrial pressure, AoP the aortic pressure, RAP the right atrial pressure, CP: Gain for compensation, t the time interval of aortic pressure measurement (2 sec) and τ the time constant.

appeared. Polyurethanes (Pellethane®, Angioflex®, and TM-3®) and polyurethane derivatives (F-PU® (fluoride PU) and KP-13® (polyurethane with polydimethylsyloxane)) have not only better anti-thrombogenicity but also better durability and mechanical characteristics than Biomer® and Avcothane®. Many companies that produce blood contact medical materials and devices are applying these kinds of materials and using them to coat blood-contacting surfaces.(22) Today, we do not need to use any anticoagulant drugs for total artificial heart animals and require a slight elongation of clotting time for clinical applications (Fig. 8).

Even in the natural heart cavity, blood coagulates and thrombi form at blood stagnation points. Flow visualization and streamline analysis in blood handling parts, to avoid such stagnation, became an essential analytical method for designing artificial heart pumps and many basic

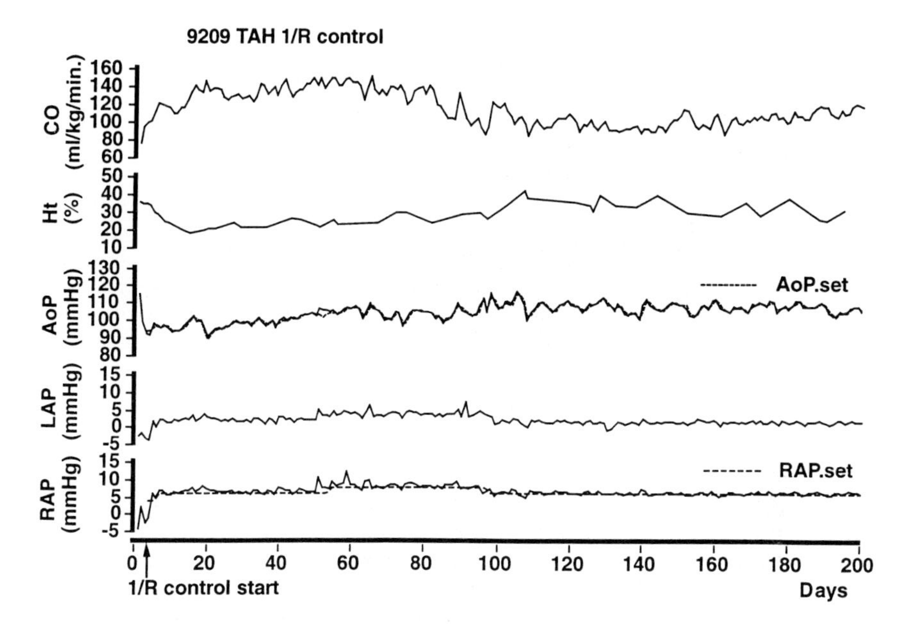

Fig. 5 A data chart for the longest surviving artificial heart transplant goat, which was controlled with an automatic feedback driving system. She survived for 360 days. (Adapted from ref. 23.) CO, left heart pump output; Ht, haematocrit; AoP, aortic blood pressure; LAP, left atrial pressure; RAP, right atrial pressure. Dotted lines indicated computer setting values.

fluid dynamic scientists joined the research. Design for a grooveless blood pump requires seamless fabrication techniques or flat surface jointing techniques. To date, surface design has together with the antithrombogenicity of lining material, been the most important factor in determining length of survival.

A total artificial heart has four valves. Recent total artificial hearts use commercially available heart valve prostheses. Most of the valves are mechanical and expensive. The dynamic characteristics of the valve are poorer than those of natural heart valves. K. Imachi has developed a polymer valve named 'jellyfish valve' which moves like a jellyfish in fluid. The valve performs better than other mechanical valves and clot formation is rare. The valve has been installed in many artificial heart pumps (Fig. 9).

Energy conversion systems

In the early 1980s, artificial heart research met a turning point, and research directions seem to divide two ways. One was the project for practical clinical application. The implantable, tether-free, LVAD for tem-

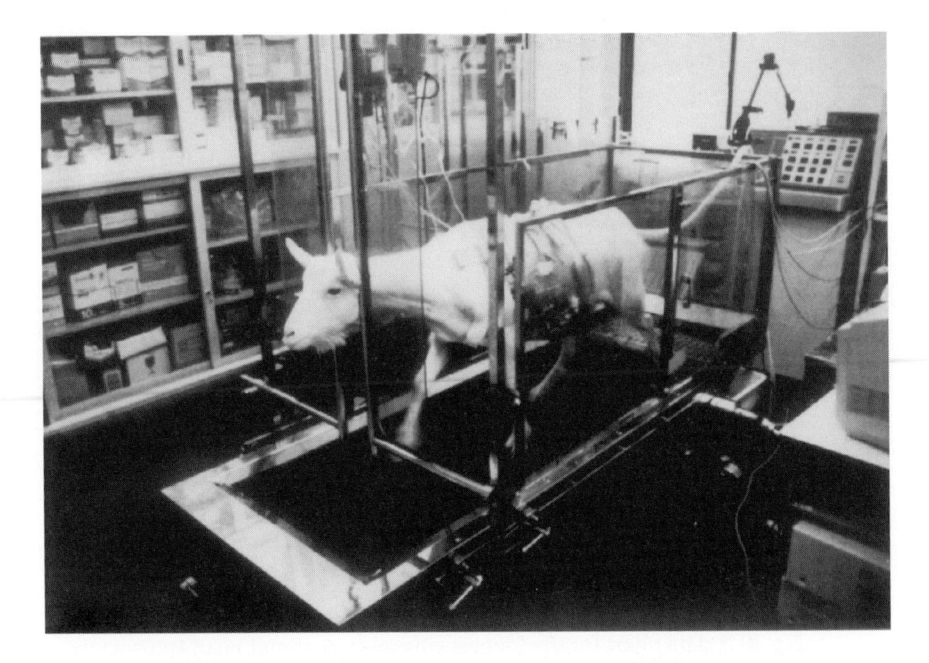

Fig. 6 An artificial heart goat running on a treadmill.

porary use became the most important target. The driving systems are moving on from pneumatic to electro-magnetic actuation. The operation procedures are changing from total cardiac replacement to left ventricular bypass operation, by connecting a total bypass pump from the left ventricular apex to the aorta and placing it subcutaneously in the abdominal wall. Transforming electric energy through intact skin, implantable battery packs for temporary driving sources, and dissipation of wasted heat from an assistance device are practical items for development today. But, in the market, many kinds of LVADs have been applied clinically. This is a completely industrialized project.

The other research direction is towards a total implantable artificial heart for permanent use. In the USA, NHLBI guided the total artificial heart system project in quite the same way as the LVAD systems. The project also aimed at temporary use and practical approach which means assembling conventional electronic parts. It is usually said that simple systems are the best. This implies that the artificial heart function is fixed at the least and the most basic requirement and many residual cardiac functions depend upon adaptability of the living body. But the target is still very difficult to achieve today.

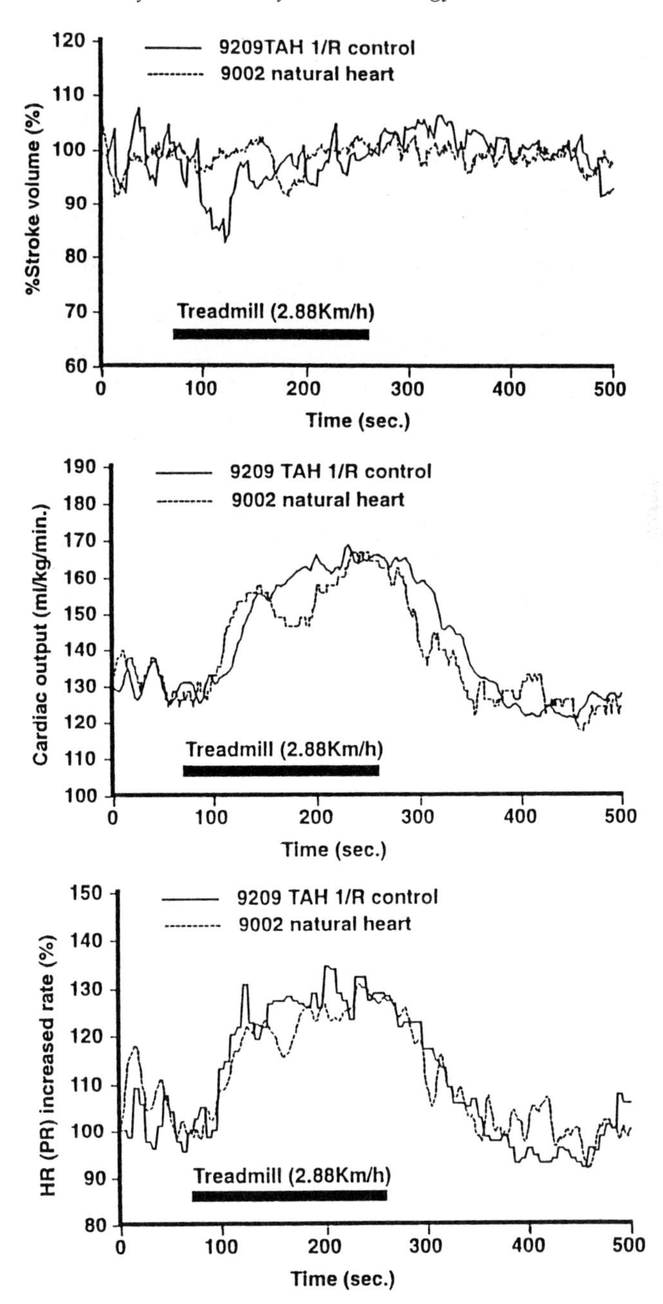

Fig. 7 The effect of exercise (running on a treadmill) on a goat with a natural heart and another with an artificial heart (TAH) controlled with the $1/R$ feedback function written in Fig. 4. Three parameters were assessed: stroke volume (top), cardiac output (middle), and heart or pulse rate [HR(PR)]. The curves for the two goats were similar in all cases. (Adapted from ref. 23.)

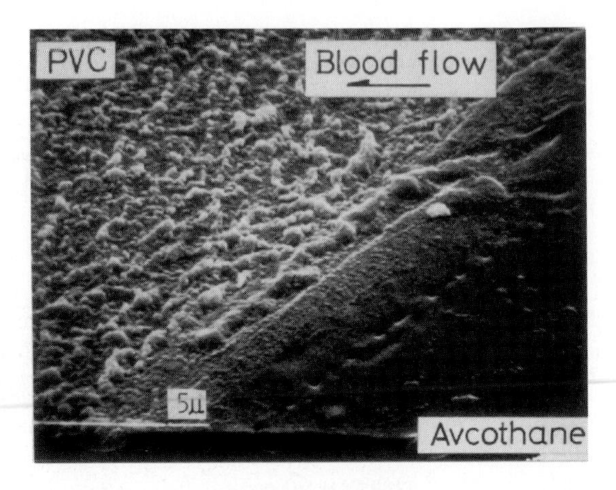

Fig. 8 Difference in clot formation on a cannula surface. No clots formed on the Cardiothane® (formerly Avcothane®) coated surface, but thick clots formed on the polyvinyl chloride coated surface. (Adapted from ref. 22.)

New driving systems

Despite our dissatisfaction regarding the delay of the practical clinical application for total, permanent artificial heart systems, there have been many spin-offs from artificial heart research in science and engineering. New driving systems, new concepts of disseminated artificial heart systems, new control methods, and new actuators are seeds for future artificial heart technology which are based upon the new science arising from past research.

Thirty years after artificial heart research first started, researchers have developed many kinds of artificial heart driving systems. Engineers are liable to think that we have almost exhausted all possible types of driving systems. But it is often at just such a time when new ideas arise.

Precessional displacement pump Instead of centrifugal or turbine pumps of high efficiency that produce non-physiological continuous flow, we developed a precessional displacement pump (PDP-AH) that squeezes continuously.[24] The pump consists of a tilted disk which rotates precessionally around a pivoted axis and pushes blood continuously. It has a much shorter approach running time than centrifugal or turbine pumps (see the structure in Fig. 10). The pump produces completely pulsatile flow when the pump axis is rotated pulsatively with an electric motor. The pump is the smallest artificial heart pump today (Fig. 11).

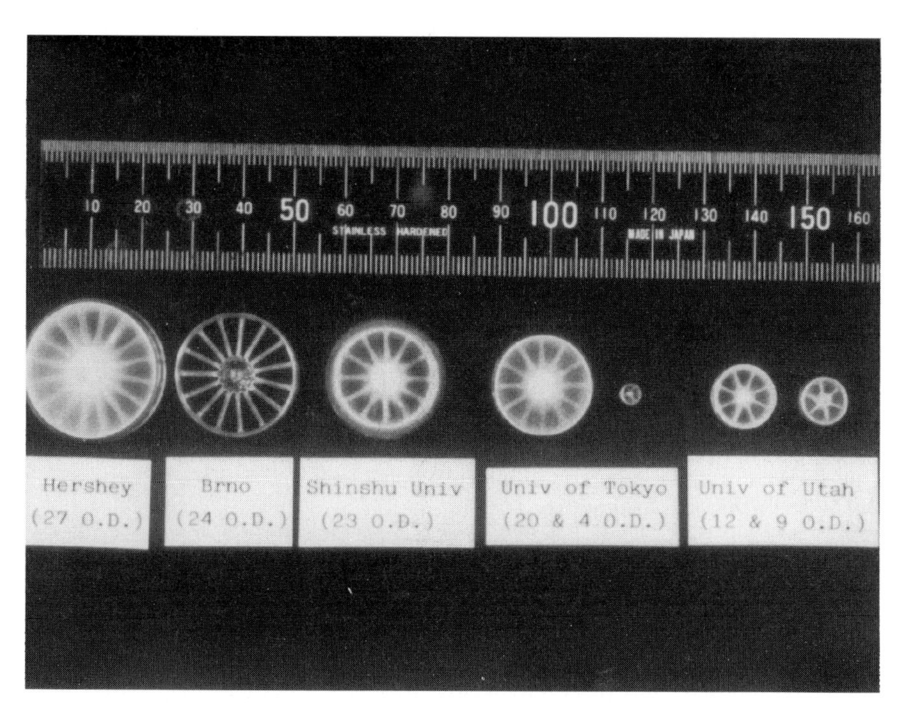

Fig. 9 'Jellyfish' valves: a new polymer membrane valve for an artificial heart pump (invented by K. Imachi).

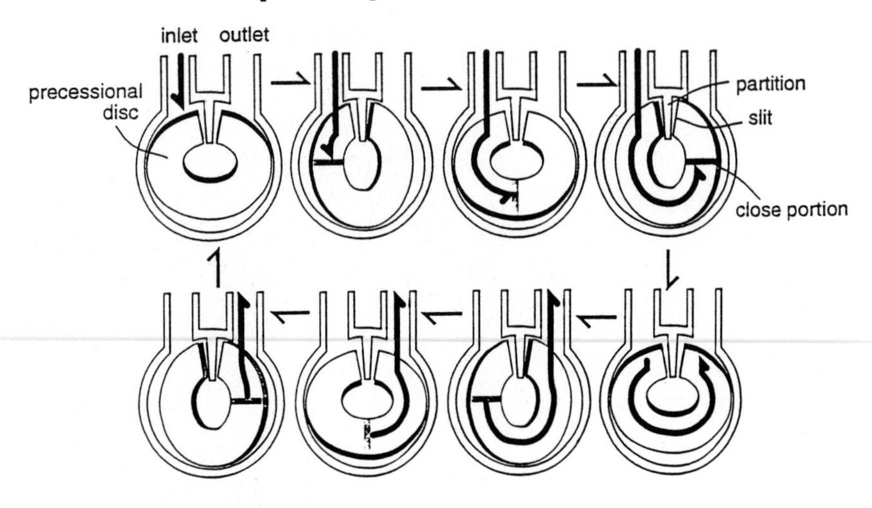

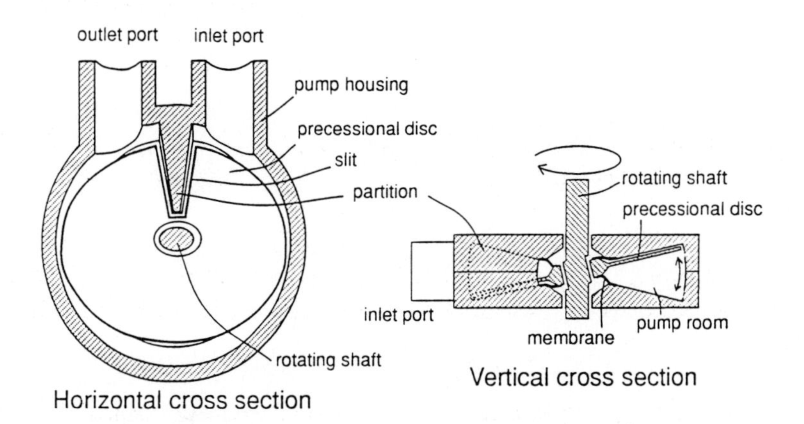

Fig. 10 A precessional displacement pump artificial heart (PDP-AH) composed of a tilted disk with a slit in a pump housing, which is covered with sealing membrane on both sides. The disk rotates precessionally around a pivoted axis and pushes blood continuously. (Developed by Y. Abe.)

Flow-transformed pulsatile total artificial heart Past research has shown that pulsatile blood flow is not essential for human life, but it is the normal physiological status. The continuous flow pump has high energy efficiency, and the continuous flow is periodically switched from systemic circulation and to pulmonary circulation. A flow-transformed pulsatile total artificial heart (FTPT-AH) consists of a single continuous

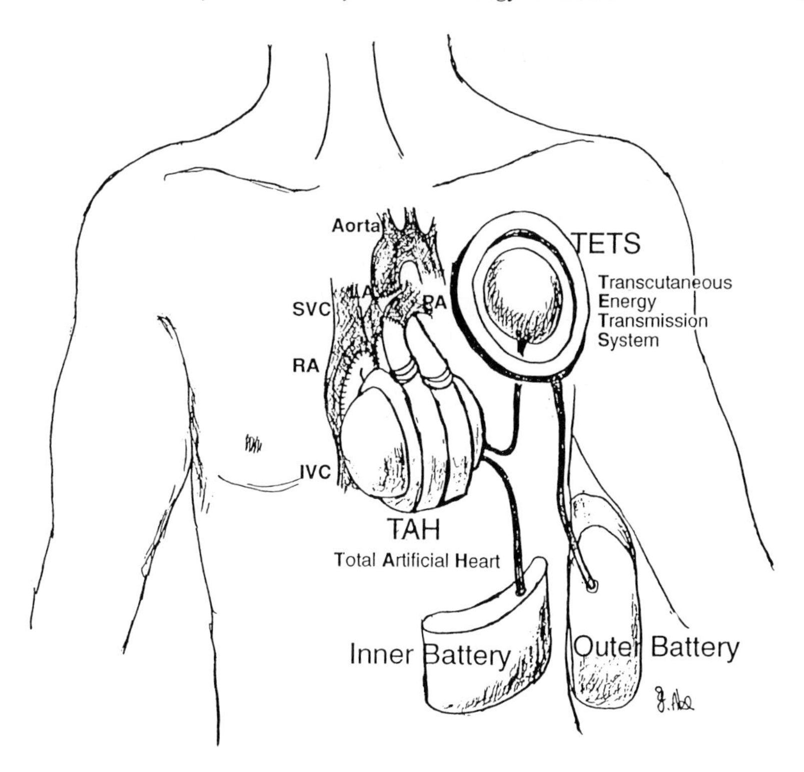

Fig. 11 The near future: a totally implantable artificial heart system using PDP-AH.

flow pump and two three-way valves and can perfuse the pulmonary circulation and the systemic circulation alternately with pulsatile flow (Fig. 12).[25]

Free diaphragm type pump In the two artificial hearts discussed above, the volume of the pump's outer shell does not change throughout the heart cycle. When the system includes an energy source and automatic controlling unit, it is self-contained and no other accessories have to be implanted in the body. But the most practical electrically driven pusher-plate type of artificial heart requires a large volume compensator usually called a compliance chamber (or bag). Ventricular displacement causes large pressure resistance in the hermetically sealed driving systems without the compliance chamber. The bag is usually the same volume as two ventricles and is installed in the chest cavity. We usually want to avoid using such a volume compensator. A fluid-driven heart can compensate for the volume outside of the body or driving system itself. A free diaphragm type pump (FDP) has a pusher-plate that includes a one-way

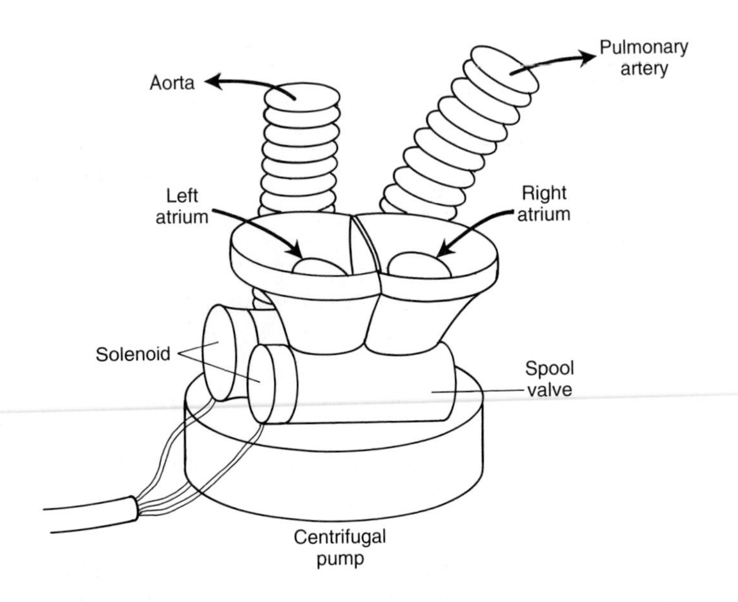

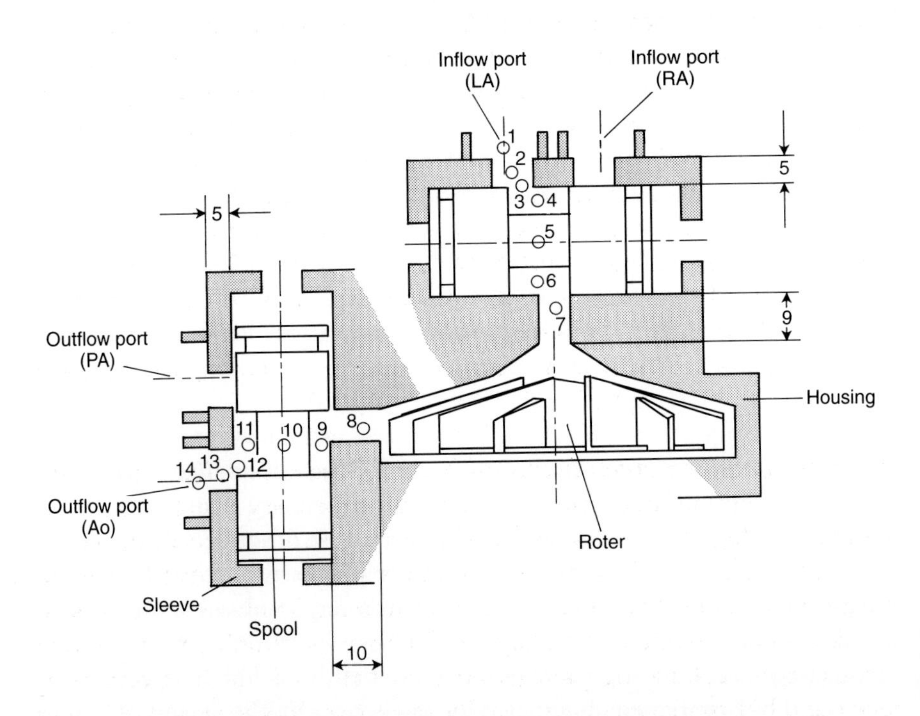

Fig. 12 A flow transformed pulsatile total artificial heart (FTPT-AH) consisting of a single continuous flow pump and two three-way valves. It can perfuse the pulmonary and systemic circulation alternately with pulsatile flow. (Developed by T. Isoyama.)

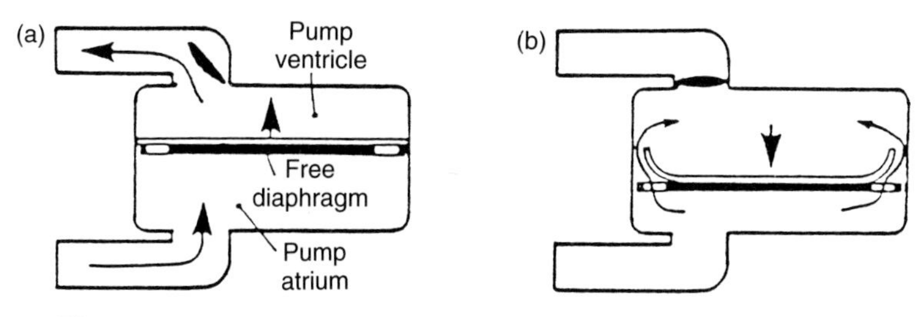

Fig. 13 Precessional displacement (continuous flow displacement). (a) Systole; (b) diastole.

valve on itself and does not require any compliance bag (Fig. 13).[26,27] The pump was applied to a disseminated artificial heart system actuated by a miniature linear electric motor.

Dispersed autonomous artificial heart system Some critics speak about the essential requirement of a permanent artificial heart system. They hope such a heart should last for at least five years, maintain the performance of the original heart over this period, and assure the recipient's quality of life. They add one more request; the heart should not stop suddenly but gradually. They fear that the electric artificial heart might stop suddenly without warning. The anxiety indicates that engineers should develop multiple actuator systems. The disseminated artificial heart system is one solution to this problem. If we can develop a miniature artificial heart system which can be tested in a rabbit or a mouse, we would be able to install many heart pumps near the important organs.[28] If each heart pump controls the output autonomously and works with the others in systematically close connection, we will have achieved a new concept of artificial heart (Fig. 14).

Artificial muscle driven artificial heart Fundamentally, if we could develop a completely muscle-compatible actuator that consisted of a great number of integrated microactuator elements, got energy inside the body, and contracted more than 20 per cent of the length, we would be able to design the ideal artificial heart (Fig. 15). Artificial heart researchers have tested some shape memory alloys and polymers which can memorize the shape of heat setting piezo-electric crystals, and chemo-mechanical materials, but these materials usually have low power output and low frequency responses. Since 1987, micromechanical system projects have started in the USA, Japan, Germany, and the Netherlands. These projects, usually called 'micromachine projects', include development of new microactuators. An electrostatic linear film actuator seems a good example of

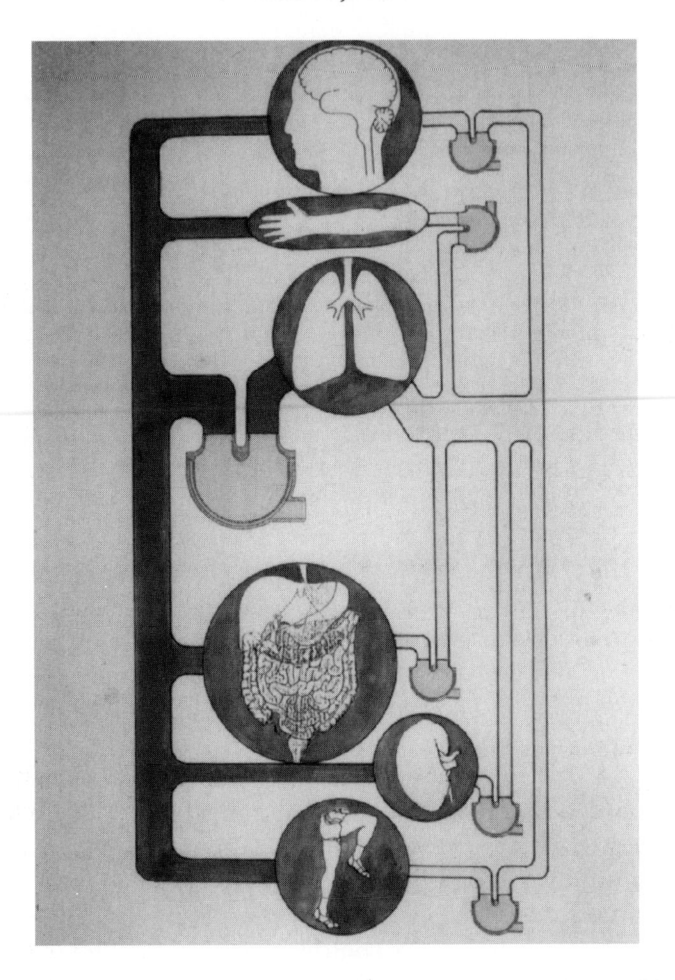

Fig. 14 Decentralized systemic blood pump system (designed by K. Imachi).

an artificial muscle element.[29] We started artificial muscle research from the standpoint of artificial structured material based upon mimicking the structure of striated muscle.[30] The research led to the development of a new *genre* in mechanical engineering and biophysics because mechanical and cellular dynamics in the size range 1μm to 1 nm had not been analysed before. This field is called mesoscopic mechanics.

Future prospects

In 2007, we shall celebrate 50 years of artificial heart research. Will we by then have implanted a totally implantable and tether-free artificial

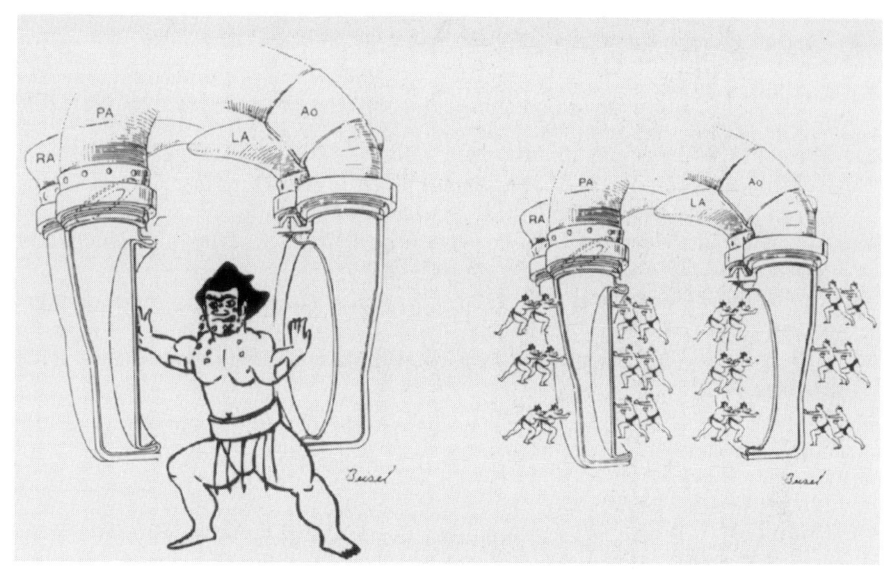

Fig. 15 A blood pump driven by an integrated micro-actuator system (designed by K. Imachi).

heart system in humans? If cardiac surgeons can get an artificial heart which is compatible with the natural heart and will last for five years, they will begin to see artificial heart replacement as a viable alternative to heart transplantation. Artificial heart implantation may become as common as fitting a pacemaker. The time will come when a truly mechanical heart can be implanted in humans.

References

1. DeVries, W. C. (1982). *Trans. Am. Soc. Artif. Intern. Organs*, **28,** 652–5.
2. Akutsu, T. and Kolff, W. J. (1958) *Trans. Am. Soc. Artif. Intern. Organs*, **4,** 230–2.
3. Frazier, O. H., Akutsu, T., and Colley, D. A. (1982). *Trans. Am. Soc. Intern. Artif. Organs*, **28,** 534–42.
4. DeWall, R. A. *et al.* (1956). *Surg. Clin. N. Amer.* **36,** 1025–31.
5. Fujimasa, I., Imachi, K., Nakajima, M. *et al.* (1986). In *Progress in artificial Organs—1985* (ed. Nose, Y., Kjelletrand, C., and Ivanovich, P.), pp. 345–53. ISAO Press, Cleveland, CH.
6. Aufiero, T. X., Magovern, J. A., Rosenberg, G. *et al.* (1987) *Trans. Am. Soc. Artif. Intern, Organs*, **33,** 157–61.
7. Taylor, K. D., Gaykowski, R., Keate, K. S. *et al.* (1987) *Trans. Am. Soc. Artif. Intern. Organs*, **33,** 738–43.
8. Kunin, C. M., Dobbins, J. J., Melo, J. C. *et al.* (1988). *J. Am. Med. Assoc.*, **259,** 860–5.

9. Snyder, A., Rosenberg, G., Weiss, W. *et al.* (1991). *Trans. Am. Soc. Artif. Intern. Organs*, **37,** M237.
10. McGee, M. G., Parnis, S. M., Nakatani, T. *et al.* (1989). *Trans. Am. Soc. Artif. Intern. Organs*, **35,** 614–16.
11. Portner, P. M., Oyer, P. E., Miller, P. J. *et al.* (1978). *Artif. Organs*, **2,** 402–12.
12. Golding, L. R., Jacobs, G. B., Murakami, T. *et al.* (1980). *Trans. Am. Soc. Artif. Intern. Organs*, **26,** 251–5.
13. Acker, M. A., Hammond, R. L., Mannion, J. D. *et al.* (1987). *Science*, **236,** 324–7.
14. Kochamba, G. and Chui, C.-J. (1987). *Trans. Am. Soc. Artif. Intern. Organs*, **33,** 404–7.
15. Hiller, K. W., W., S., and Kolff, W. J. (1962). *Trans. Am. Soc. Intern. Artif. Organs*, **8,** 125–30.
16. Imachi, K., Fujimasa, I., Oomichi, H. *et al.* (1976). *Jinkozoki (Artif. Organs)*, **5,** 321–4.
17. Joyce, L. D., DeVries, W. C., Hastings, W. L. *et al.* (1983) *Trans. Am. Soc. Artif. Intern. Organs*, **24,** 81–7.
18. Landis, D. L., Pierce, W. P., Rosenberg, G. *et al.* (1977). *Trans. Am. Soc. Artif. Intern. Organs*, **23,** 519–25.
19. Klain, M., Murava, G. L., Tajima, K. *et al.* (1971). *Trans. Am. Soc. Artif. Intern. Organs*, **17,** 437–45.
20. Broetos, J. W., Pierce, W. S., and Baier, R. E. (1975). *J. Biomed. Mater. Res.*, **9,** 327–40.
21. Nyilas, E. J. (1972). *Biomed. Mater. Res.* (Symp.) **3,** 97–127.
22. Imachi, K., Mabuchi, K., Chinzei, T. and Fujimasa, I. (1993). In *New functionality materials* (ed. T. Tsuruta, M. Doyama, M. Seno and Y. Imanishi), pp. 353–56. Elsevier, Amsterdam.
23. Abe, Y. (1994). *A physiological control method of a total artificial heart.* The University of Tokyo.
24. Abe, Y., Chinzei, T., Isoyama, T., Imachi, K. *et al.* (1993). *Artif. Organs*, **17,** 523.
25. Imachi, K., Isoyama, T., Chinzei, T., Abe, Y. *et al.* (1992). *Trans. Am. Soc. Artif. Intern. Organs*, **38,** M717.
26. Atsumi, K., Imachi, K., Fujimasa, I., and Sezai, Y. J. (1990). *Biomat. Appl.* **4,** 161–224.
27. Imachi, K., Chinzei, T., Abe, Y., Mabuchi, K. *et al.* (1991). In *Artificial heart 3*, pp. 137–42 (ed. Akutsu. T. and Koyanagi, H.). Springer, Tokyo.
28. Imachi, K., Chinzei, T., Abe, Y., Mabuchi, K. *et al.* (1993). *Artif. Organs*, **17,** 525.
29. Egawa, S. and Higuchi, T. (1990). *Multi-layered electrostatic film actuator. 1-166-171.* IEEE Press, Napa Valley, CA.
30. Fujimasa, I. (1993). *Artif. Organs*, **17,** 433.

IWAO FUJIMASA

Born 1937, educated at the Faculty of Medicine, University of Tokyo, where he developed artificial heart systems as a Research Associate in the Institute of Medical Electronics from 1965. Became Associate Professor 1975, and Professor 1988. 1975–78 conducted analysis on healthier systems as Research Fellow, International Institute for Applied Systems Analysis in Austria, and since 1980 has been Professor, School of Policy Science, Saitama University. Since 1986, has been Professor at the Research Center for Advanced Science and Technology, University of Tokyo, promoting a micromachine project with the Ministry of International Trade and Industry. Publications include: *Engineering for Artificial Organs*, 1988; *Physiological Function Imaging: Thermography*, 1989; *A Notebook for Micromachine Research and Development*, 1991; *Invisible Machines: Mesoscopic Engineering*, 1993.

Is high technology medicine cost-effective?

ALAN MAYNARD

Introduction

Imagine you are in a health care supermarket with, if you are British, £37bn to spend (6 per cent of Gross Domestic Product (GDP)) and, if you are American $850bn (14.5 per cent of GDP), what would you buy?

There is no shortage of advice. One group of 'actors' are the benefit maximizers who advocate investments in technology often on the basis of beliefs, rather than good evidence, about benefits, and regardless of costs. Outside medicine this group has induced investments in Concorde and the Humber bridge: glorious achievements of engineering but inefficient! Inside medicine this group has induced investments in unproved diagnostic procedures and the inappropriate use of them and pharmaceutical and other therapies. The success of this group leads to excess supply of facilities: for example, Beijing has more magnetic resonance imaging (MRI) and CT scanners than England.

This group of benefit maximizers is opposed by those miserable people, accountants! Whilst doctors know (or claim to know) the benefits of everything and know the cost of nothing, the accountants know (or claim to know) the cost of some things but are ignorant of the benefits of everything. The cheapest procedure in medicine is to do nothing and sometimes this is the most efficient. Such a policy delights the accountant!

If society's goal is to use resources efficiently (i.e. to produce additional improvements in the length and quality of patients at least cost), the measurement of benefits should be brought together with the measurement of cost. It is the role of the economist to bring the benefit maximizers and the cost minimizers together to ensure that resources are targeted to produce maximum 'health gains'.

Economics, described as the 'dismal science' in the nineteenth century, when applied to health care is concerned with harsh choices about who will life in what degree of pain and discomfort and who will die. It is impossible to meet all health care needs: resources have to be targeted at those patients who will get the greatest enhancement in the length and quality of their lives per unit of cost. Such prioritization is unpopular but happens every day in all public and private health care systems. As a consequence of this rationing, patients are denied treatments which have benefits by being made to queue sometimes for eventual treatment and sometimes for the great waiting list in the sky!

The forces of demographic change (the 'greying' of the population) and technological development are further complicating the processes of prioritization or rationing in health care. At worst this means that patients, particularly the elderly, may die in a storm of high technology medicine. The problem with such an exit from life, and indeed with all medical practice, is that the basis for deploying health technologies is not well founded in scientific knowledge.

Ideally, when clinicians use scarce resources to diagnose and treat their patients, they should be well informed about the effects (in terms of enhancements in the length and quality of life) and costs of their interventions. Unfortunately, medical practice is not knowledge-based. The purpose of this article is to explore the issue of cost-effectiveness in the use of high technology medicine. In the first section the issue of the objectives of the health care system will be explored further. It will then be shown that efficient choices in health care are frustrated by ignorance and perverse incentives. Some ways in this these problems can be mitigated and greater understanding of the cost-effectiveness of high technology medicine achieved are then explored. The concluding section offers optimistic and pessimistic scenarios for the future: is greater willingness to question clinical judgement emerging in Whitehall and in the population, or is the unproven use of society's scarce resources in health care likely to remain unchallenged?

Making choices in health care: what are the objectives?

Given that resources are scarce and harsh choices about who will receive what treatment when are inevitable, what objectives are to be pursued by decision makers in charge of resource allocation?

Access to health care can be rationed by the price mechanism, and determined by individuals' willingness and ability to pay, or by non-price mechanisms. Here it is assumed that the preferred alternative to the price mechanism is allocation of access based on 'need', interpreted as individual capacity to benefit per unit of cost.

Thus it is assumed that the goal of the National Health Service (NHS) is to allocate resources to patients who get the greatest benefit (i.e. increases in the length and quality of life) per unit cost. The identification of benefit and cost (i.e. cost-effectiveness) is a technical problem. The decision about the level of expenditure is a political issue resolved by Parliament deciding how much it will spend to buy an additional unit of benefit (for example, a quality-adjusted life year, or QALY).

Why is choice difficult?

The choices that determine resource allocation are difficult because of the failure to measure costs and outcomes (benefits) systematically in all areas of medical practice. This failure manifests itself in four ways, discussed in the following sections.

Ignorance about costs

The recent NHS reforms have led to the production of some price data. However, these price data exhibit large variations for the same procedures and are the product of diverse accounting practices which allocate overheads in different ways as well as differences in treatment patterns. These prices are not measures of the opportunity cost (i.e. the value of what is foregone when a procedure is provided) of a patient episode of care. Most decisions in the NHS and other European health care systems are made in the absence of cost data.

Ignorance about outcomes

There is ignorance not only about the value of what is given up (i.e. the cost) when treatment is provided, but also of the value of the benefit of health care. For instance, there is little measurement of a hospital's, let alone a clinician's success in terms of in-patient mortality. This is despite sustained advocacy over many centuries. For instance, Francis Clifton, physician to George II in 1732, argued:

> In order, therefore to procure this valuable collection, I humbly propose, first of all, that three or four persons should be employed in the hospitals (and that without any ways interfering with the gentlemen now concerned), to set down the cases of the patients there from day to day, candidly and judiciously,

> without any regard to private opinions or public systems, and at the year's end publish these facts as they are, leaving every one to make the best use he can for himself.
>
> Francis Clifton, quoted in a *Lancet* editorial, 1841 pp, 650–651)

This argument was reiterated more recently by Florence Nightingale who advocated that a hospital's success be measured in terms of whether a patient was dead, relieved, or unrelieved. She went on to argue,

> I am fain to sum up with an urgent appeal for adopting this or some uniform system of publishing the statistical records of hospitals. There is a growing conviction that in all hospitals, even in those which are best conducted, there is a great and unnecessary waste of life . . . In attempting to arrive at the truth, I have applied everywhere for information, but in scarcely an instance have I been able to obtain hospital records fit for any purpose of comparison. If they could be obtained, they would enable us to decide many other questions besides the ones alluded to. They would show subscribers how their money was being spent, what amount of good was really being done with it, or whether the money was doing mischief rather than good.
>
> Florence Nightingale (1863)

By the twentieth century Americans too were advocating the measurement of hospitals' success:

> We must formulate some method of hospital reporting showing as nearly as possible what are the results of the treatment obtained at different institutions. This report must be made out and published by each hospital in a uniform manner, so that comparison will be possible. With such a report as a starting point, those interested can begin to ask questions as to management and efficiency.
>
> E. A. Codman (1914)

Dr Codman was, for raising these issues, tarred, feathered, and run out of town.

By the 1980s the bold were demanding 'value for money' and Reagan's government began to publish US mortality data by hospital in 1987. Similar data were published in the UK in 1988 by Kind (Table 1).[2]

Are these data useful? The differences, adjusted for age and sex, may be due to factors such as case severity, measurement inaccuracies, and the socio-economic background as well as the hospital's and the doctor's success, so they have to be used with care. But is it not curious that these data had been collected in the UK for 20 years but never evaluated or used to inform both clinical and non-clinical managerial review?

Of course, in-patient survival data need to be augmented with data

Table 1. Do some hospitals 'kill' more patients than others?

District	Crude mortality rate (%)	Standard mortality rate
North Tees	6.4	1.317
Grimsby	6.4	1.414
Scarborough	6.9	1.214
Sheffield	4.7	0.829
Harrow	3.4	0.577
North-west Hertfordshire	5.9	1.510
Brighton	9.3	1.500
Macclesfield	8.8	1.330
Central Manchester	2.8	0.695

Source: ref. 2, Table 4.

regarding survival in the longer term (e.g. over 5, 10, or 20 years) and the quality of survival. Does the patient recover fully after a procedure or is she left paralysed or unconscious? If benefit is to be measured such data need to be collected to identify improvements in the length and quality of life produced by competing procedures.

Variations in medical practice

Given both the failure to measure costs and effects systematically and the uncertainty associated with much medical practice, it is not surprising to see large variations in medical practice. Practitioners do very different things to patients of similar age, sex, and medical problem (Table 2).

What explanation of these variations have been discussed in the literature? The variations which exist between doctors, districts, regions, and countries are affected by demand (resourcing), supply (availability), morbidity (illness), and clinical factors. The clinical causes of the variations in practice are large and reflect the absence of agreement about appropriate therapies and the absence of clinical guidelines both to guide practice and to make practitioners accountable.

Ignoring effectiveness data

The evaluation of the effectiveness of medical interventions is modest but useful. Using its products is difficult because the quality of these data varies and inadequate trial design and execution can produce biased data and results which misinform practice. The best sort of data comes from well designed randomized controlled trials. However, choices can also be informed by other types of trials (such as case-controlled trials) and expert

Table 2. Magnitude of systematic variation (in ascending order) for selected causes of admission among 30 hospital market areas in Maine: 1980–1982

Variation	Medical	Surgical
Low: 1.5-fold range	–	Inguinal hernia repair Hip repair
Moderate: 2.5-fold range	Acute myocardial infarction Gastrointestinal haemorrhage Cerebrovascular accident	Appendectomy Major bowel surgery Cholecystectomy
High: 3.5-fold range	Respiratory neoplasms Cardiac arrhythmias Angina pectoris Psychosis Depressive neurosis Medical back problems Digestive malignancy Adult diabetes	Hysterectomy Major cardiovascular operations Lens operations Major joint operations Anal operations Back and neck operations
Very high: ⩾8.5-fold	Adult bronchioloitis Chest pain Transient ischaemic attacks Minor skin disorders Chronic obstructive lung disease Hypertension Atherosclerosis Chemotherapy	Knee operations Transuretheral operations Extraocular operations Breast biopsy Dilatation and curettage Tonsillectomy Tubal interruption

(From ref. 12.)

judgements made, for instance by consensus conferences. The quality of such information is less robust. As Abba Eban (the former Israeli foreign minister) remarked, 'Consensus means that lots of people say collectively what nobody believes individually.'

Where data are available, they can inform choices. Examples are given below.

1. The majority of children with 'glue ear' could be treated best by 'watchful waiting', since the problem resolves itself, rather than by surgical procedures to insert grommets.[3]
2. The diagnostic use of dilatation and curettage (at six times the level used in the USA) is inappropriate and a waste of resources for most young women.[4]
3. The efficacy of expensive SSRIs and tricyclics in the treatment of depression is equal, as is compliance, and as the latter are up to 30 times cheaper they should be used.[5,6]

4. The use of lipid-lowering drugs to treat high cholesterol levels should be targeted at high risk patients if these expensive drugs are to be used effectively.[7,8]

However, these data often tend to be ignored: it is difficult to make practitioners change their methods.[9]

Similarly, it is difficult to get these same doctors and managers to recognize and react to the issues of economies of scale i.e. reduced mortality created by specialization and the treatment of large volumes of patients. It seems that surgeons who treat more aortic aneurysms have lower patient death rates. It also seems that those centres which carry out more than 80–100 kidney transplants get better outcomes. If practitioners specialize and if specialist centres are created (e.g. in cancer chemotherapy and radiotherapy), outcomes may be better. The quality of evidence for such conclusions is indicative rather than unambiguous, and merits more careful investigation and appropriate management action.

Conclusions

Choices are difficult to make because most technologies have not been evaluated. Fuchs argued that 10 per cent of health care expenditure reduces patients' health, 10 per cent has no effect, and 80 per cent of expenditure on health care improves health. [10] The problem, he noted, is that we do not know which therapies lie in each per cent category!

Thus the problems involved in investigating the cost-effectiveness of high technology, and indeed all, medicine, are that we do not know what works in the majority of treatments and, where we do know what works, we do not know how to change practitioners' behaviour. How can we do better?

Improving resource allocation

Choices are made

Choices are made in every health care system at every hour of the day. They tend to be implicit, inconsistent, and incomprehensible. The response of some policy makers is to make the processes of resource allocation explicit and more consistent.

For instance in Oregon, USA, the State Government set up a Commission to prioritize some 700 procedures in terms of their effectiveness in order to ration care amongst Medicaid recipients (the poor). They analysed usage and cost data to decide a budget and draw a line in their 'league table' below which procedures would not be financed. This was a crude but explicit process which reflected judgements more than systematic measurement of cost-effectiveness. Despite such caveats as these and the aggregate nature of the procedures, President Clinton has agreed to an experiment with the scheme, although it may be lost in the hiatus created by Hilary Clinton's reform proposals.

In Holland the Dunning Committee[11] outlined a process for determining priorities: the identification of necessary, then efficacious, then effective, and then socially desirable care. These four stages, or filters, the Committee concluded should be used to determine a core package of proven health care benefits that the State would finance. The Dutch have yet to move from these general principles to practice. However, similar discussions are taking place in Sweden (with a report likely in November, 1993) and in Germany, where a discussion on the 'core' package of State health services is likely in 1994. The British are inefficiently leaving these processes of prioritization to local purchasers: what is needed here too is a national debate about the criteria to be used and how such principles might be translated into practice.

Measuring cost-effectiveness

Such prioritization must be better informed by the scientific measurement of the costs and effects of competing interventions. This is not an easy task. One difficulty is that the typical clinical trial design tends to use narrow clinical end-points and to ignore issues related to the measurement of cost and quality of survival. The appropriate clinical trial design is one which facilitates the measurement of cost-effectiveness and this requires multi-disciplinary teams to design and execute protocols.

Such protocols might incorporate a variety of techniques of economic evaluation. If the outcomes of the alternatives are known from existing effectiveness studies to be identical, the appropriate approach is cost minimization. If the outcomes of the alternatives differ then it may be appropriate to use other techniques (Table 3), especially cost-effectiveness analysis.

As now, the results of such trials may not have any impact on practice. To ensure that this does not happen, the incentive structure has to change. For instance, in Australia new pharmaceuticals are evaluated for safety and efficacy to get a product licence. This enables the company to sell the drug. However, reimbursement in the public health care system is now determined by a separate process which requires that companies demonstrate that the drug is cost-effective. If they fail to do this, they can sell the drug but it will not be reimbursed by the State system which funds

Table 3. Types of economic evaluation

	Cost measurement	Outcome measurement: what	Units used to express outcome
Costing	Pounds	Assumed identical	None
Cost–benefit analysis	Pounds	All effects produced by the alternative	Pounds
Cost-effectiveness analysis	Pounds	Single common specific variable achieved to varying extents	Common units (e.g. life years)
Cost–utility analysis	Pounds	Effects of the competing therapies and achieved to differing levels	QALYs or HYEs

most care. This mechanism has transformed the industry's interests in economic evaluation.

Conclusions

Is high technology medicine cost-effective? We do not know the answer: doctors have exploited their freedom to innovate and use interventions, most of which are unproven in terms of effectiveness, let alone cost-effectiveness. If such clinical freedom produces inefficiency in the use of scarce resources it may be unethical: inefficiency wastes resources which could be used to treat patients (e.g. those on a waiting list) who would benefit from care.

Will this change? An optimistic view is that change is happening with the power of providers (clinical, pharmaceutical, and equipment providers) being challenged. Purchasers of health care, after challenging practice, are seeing more clearly both that often the 'provider Emperor' has no clothes and that there is a desperate need for systematic and scientific health services research.

The pessimistic view about change is that it is not happening. Policy makers often respond to crises in health care by 'redisorganizing' the structure of the system rather than by confronting and resolving the fundamental problems inherent in medical practice. In the UK the NHS is reorganized every four or five years and the architects of these reforms usually fail to address fundamental issues or, if they do, the effects are diluted if not dissipated by provider power. This outcome is familiar for anyone who works in health and social care and, as an economist and a 'dismal scientist' it is one on which I will conclude with a quotation from an administrator who worked for the Roman Emperor Nero.

> We trained very hard, but it seemed that every time we were beginning to form up into teams, we would be reorganised. I was to learn later in life that we tend to meet any new situation by reorganising, and a wonderful method it can be for creating the illusion of progress, while producing confusion, inefficiency and demoralisation.
>
> Caius Petronius (AD 66)

References

1. Codman, E. A. (1914). *Surg. Gynaecol. Obstet.*, **18,** 494.
2. Kind, P. (1988). *Hospital deaths—the missing link: measuring outcomes in hospital activity data.* Discussion Paper no. 44, Centre for Health Economics, University of York.
3. Freemantle, N. *et al.* (1992). The treatment of persistent glue ear. *Effective Health Care*, Bulletin No. 4. Centre for Health Economics (York) and School of Public Health (Leeds).
4. Coulter, A., Klassen, A., MacKensie, I. Z., and McPherson, K. (1993). *Brit. Med. J.*, **306,** 236–9.
5. Song, F., Freemantle, N., Sheldon, T. A., *et al.* (1993). *Brit. Med. J.*, **1215,** 683–7.
6. Freemantle, N. *et al.* (1993). The treatment of depression in primary care. *Effective Health Care*, Bulletin No. 5. Centre for Health Economics, (York) and School of Public Health (Leeds).
7. Davey-Smith G., Song, F., and Sheldon, T. A. (1993). *Brit. Med. J.*, **306,** 1367–73.
8. Freemantle, N. *et al.* (1993) Cholesterol: screening and treatment. *Effective Health Care Bulletin No. 6*, School of Public Health, Leeds University.
9. Lomas, J. (1993). *Teaching old (and not so old) docs new tricks: effective ways to implement research findings.* CHEPA Working Paper Series No. 93–4, McMaster University, Hamilton, Ontario.
10. Fuchs, V. (1984). *New Engl. J. Med.*, **311** (24), 1572–3.
11. Government Committee on Choices in Health Care (1992). *Choices in health care.* Dunning Report, Ministry of Welfare, Health and Cultural Affairs, Rijswilk, Netherlands.
12. Wennberg, J. E., McPherson, K., and Caper, P. (1984). *New Engl. J. Med.*, **311,** 295–300.

ALAN MAYNARD

Born 1944, educated University of Newcastle upon Tyne and University of York. Appointed Professor of Economics and Director of the Centre for Health Economics at the University of York in 1983. Was a member of the

Economic and Social Research Council (1986–88) and the Health Services Research Committee of the Medical Research Council (1986–92). He was a member of York Health Authority (1982–92) and is now a non-executive Director of the York NHS Trust Hospital. He has worked as a consultant for the European Commission, the World Bank, and the World Health Organization and has written extensively on economic aspects of health and health care.

THE ROYAL INSTITUTION

The Royal Institution of Great Britain was founded in 1799 by Benjamin Thompson, Count Rumford. It has occupied the same premises for nearly 200 years and, in that time, a truly astounding series of scientific discoveries has been made within its walls. Rumford himself was an early and effective exponent of energy conservation. Thomas Young established the wave theory of light; Humphry Davy isolated the first alkali and alkaline earth metals, and invented the miners' lamp; Tyndall explained the flow of glaciers and was the first to measure the absorption and radiation of heat by gases and vapours; Dewar liquefied hydrogen and gave the world the vacuum flask; all who wished to learn the new science of X-ray crystallography that W. H. Bragg and his son had discovered came to the Royal Institution, while W. L. Bragg, a generation later, promoted the application of the same science to the unravelling of the structure of proteins. In the recent past the research concentrated on photochemistry under the leadership of Professor Sir George (now Lord) Porter, while the current focus of the research work is the exploration of the properties of complex materials.

Towering over all else is the work of Michael Faraday, the London bookbinder who became one of the world's greatest scientists. Faraday's discovery of electromagnetic induction laid the foundation of today's electrical industries. His magnetic laboratory, where many of his most important discoveries were made, was restored in 1972 to the form it was known to have had in 1845. A newly created museum, adjacent to the laboratory, houses a unique collection of original apparatus arranged to illustrate the more important aspects of Faraday's immense contribution to the advancement of science in his fifty year's work at the Royal Institution.

Why the Royal Institution is Unique

It provides the only forum in London where non-specialists may meet the leading scientists of our time and hear their latest discoveries explained in everyday language.

It is the only Society that is actively engaged in research, and provides lectures covering all aspects of science and technology, with membership open to all.

It houses the only independent research laboratory in London's West End (and one of the few in Britain)—the Davy Faraday Research Laboratory.

What the Royal Institution Does for Young Scientists

The Royal Institution has an extensive programme of scientific activities designed to inform and inspire young people. This programme includes lectures for primary and secondary school children, sixth form conferences, Computational Science Seminars for sixth-formers and Mathematics Masterclasses for 12–13 year-old children.

What the Royal Institution Offers to its Members

Programmes, each term, of activities including summaries of the Discourses; synopses of the Christmas Lectures and annual Record.

Evening Discourses and an associated exhibition to which guests may be invited.

An annual volume of the *Proceedings of the Royal Institution of Great Britain* containing accounts of Discourses.

Christmas Lectures to which children may be introduced.

Meetings such as the RI Discussion Evenings; Seminars of the Royal Institution Centre for the History of Science and Technology, and other specialist research discussions.

Use of the Libraries and borrowing of the books. The Library is open from 9 a.m. to 9 p.m. on weekdays.

Use of the Conversation Room for social purposes.

Access to the Faraday Laboratory and Museum for themselves and guests.

Invitations to debates on matters of current concern, evening parties and lectures marking special scientific occasions.

Royal Institution publications at privileged rates.

Group visits to various scientific, historical, and other institutions of interest.

Evening Discourses

The Evening Discourses have been given regularly since 1826. They cover all aspects of science and technology (with regular ventures into the arts) in a form suitable for the interested layman, and many scientists use them to keep in touch with fields other than their own. An exhibition, on a subject relating to the Discourse, is arranged each evening, and light refreshments are available after the lecture.

Christmas Lectures

Faraday introduced a series of six Christmas Lectures for children in 1826. These are still given annually, but today they reach a much wider audience through television. Titles have included: 'The Languages of Animals' by David Attenborough, 'The Natural History of a Sunbeam' by Sir George Porter, 'The Planets' by Carl Sagan and 'Exploring Music' by Charles Taylor.

The Library

The Library contains over 40,000 volumes, and is particularly strong in long runs of scientific periodicals. It has a fine collection of the history of science and science for the non-specialist. A selection of newspapers and magazines is available in the Conversation Room which is still characteristic of a library of the early nineteenth century.

Schools' Lectures

Extending the policy of bringing science to children, the Royal Institution provides lectures throughout the year for school children of various ages, ranging from primary to sixth-form groups. These lectures, attended by thousands, play a vital part in stimulating an interest in science by means of demonstrations, many of which could not be performed in schools.

Seminars, Masterclasses, and Primary Schools' Lectures

In addition to educational activities within the Royal Institution, there is an expanding external programme of activities which are organized at venues throughout the UK. These include a range of seminars and masterclasses in the areas of mathematics, technology and, most recently, computational science. Lectures aimed at the 8–10 year-old age group are also an increasing component of our external activities.

Teachers' Workshops

Lectures to younger children are commonly accompanied by workshops for teachers which aim to explain, illustrate, and amplify the scientific principles demonstrated by the lecture.

Membership of the Royal Institution

Member

The Royal Institution welcomes all who are interested in science, no special scientific qualification being required. By becoming a Member of the Royal Institution an individual not only derives a great deal of personal benefit and enjoyment but also the satisfaction of helping to support the unique contribution made to our society by the Royal Institution.

Family Associate Subscriber

A Member may nominate one member of his or her family residing at the same address, and not being under the age of 16 (there is no upper age limit), to be a Family Associate Subscriber. Family Associate Subscribers can attend the Evening Discourses and other lectures, and use the Libraries.

Associate Subscriber

Any person between the ages of 16 and 27 may be admitted as an Associate Subscriber. Associate Subscribers can attend the Evening Discourses and other lectures, and use the Libraries.

Junior Associate

Any person between the ages of 11 and 15 may be admitted as a Junior Associate. Junior Associates can attend the Christmas Lectures and other functions, and use the Libraries. Junior Associates may take part in educational visits organized during Easter and Summer vacations.

Corporate Subscriber

Companies, firms and other bodies are invited to support the work of the Royal Institution by becoming Corporate Subscribers; such organizations make a very valuable contribution to the income of the Institution and so endorse its value to the community. Two representatives may attend the Evening Discourses and other lectures, and may use the Libraries.

College Corporate Subscriber

Senior educational establishments may become College Corporate Subscribers; this entitles two representatives to attend the Evening Discourses and other lectures, and to use the Libraries.

School Subscriber

Schools and Colleges of Education may become School Subscribers; this entitles two members of staff to attend the Evening Discourses and other lectures, and to use the Libraries.

Membership forms can be obtained from: The Membership Secretary, The Royal Institution, 21 Albemarle Street, London W1X 4BS.
Telephone: 0171 409 2992. Fax: 0171 629 3569